Proteases of Retroviruses

Proteases of Retroviruses

Proceedings of the Colloquium C 52
14th International Congress of Biochemistry

Prague, Czechoslovakia · July 10–15, 1988

Editor
Vladimír Kostka

Walter de Gruyter · Berlin · New York 1989

Editor
Vladimír Kostka, Dr., D. Sc.
Czechoslovak Academy of Sciences
Institute of Organic Chemistry and Biochemistry
Flemingovo náměstí 2
CS-166 10 Prague 6
Czechoslovakia

Library of Congress Cataloging-in-Publication Data

Proteases of retroviruses.
Proceedings of the Colloquium C 52 "Proteinases of Retroviruses."
Includes indexes.
1. Proteolytic enzymes--Congresses. 2. HIV (Viruses)--Enzymes--Congresses.
3. Retroviruses--Enzymes--Congresses.
I. Kostka, Vladimír, 1930- . II. International Congress of Biochemistry
(14th : 1988 : Prague, Czechoslovakia) III. Colloquium C 52 "Proteinases of
Retroviruses" (1988 : Prague, Czechoslovakia) [DNLM: 1. HIV-1--enzymology--
congresses. 2. Peptide Hydrolases--metabolism--congresses. 3. Retroviridae--
enzymology--congresses.
QW 166 P967 1988]
QP609.P78P748 1989 616'.0194 89-11614
ISBN 0-89925517-5 (U.S.)

Deutsche Bibliothek Cataloging-in-Publication Data

Proteases of retroviruses : proceedings, 14th International Congress of
Biochemistry, colloquium C 52, Prague, Czechoslovakia, July 10-15, 1988 /
ed. Vladimír Kostka. - Berlin ; New York : de Gruyter, 1989
ISBN 3-11-011820-3
NE: Kostka, Vladimír [Hrsg.]; International Congress of Biochemistry
<14, 1988, Praha>

⊗ Printed on acid free paper

Preface

Writing this preface I cannot resist returning in my mind to FEBS Advanced Course No 84/07, "Aspartic Proteinases and Their Inhibitors", which we organized in Prague in 1984. When the direction of future research on this topic was assessed during the closing sessions of the Course it was postulated by several speakers that attention would be focused (a) on structure-function studies at the level of the 3-D structures of aspartic proteases of mainly industrial or medical importance, then typically represented by chymosin and renin; and (b) on a search for new aspartic proteases (and acid proteases in general) in organisms belonging to the lower end of the evolutionary scale. If I remember correctly, nobody mentioned the possibility that such a search should consider retroviruses (or viruses in general, for that matter) as a starting material, even though the existence of "virus-associated proteolytic activities" in various virus systems had already been established and their role in virus maturation was more or less known. Until the mid-1980s these proteases were examined almost exclusively by virologists, mainly from this perspective, and their enzymological and molecular characteristics remained largely unknown. There is a little doubt that efforts to obtain a deeper insight into the action of these enzymes were stimulated, not only by the general trend towards molecular-level examination in virology, but also by the new concept of viral proteases as possible therapeutic targets, an idea that the AIDS threat has lent urgency to. An intensive investigation of proteases involved in the processing of picornaviruses was launched first, soon paralleled by a similarly oriented investigation of retroviruses. The discovery of simple sequence homologies between retroviral proteases and the active site sequences of aspartic proteases seemed to promise that the experience gleaned from more than fifty years of research on this "oldest" class of enzymes might be used to advantage in research on their counterparts of viral origin.

In view of these facts, the 14th International Congress of Biochemistry provided a logical platform for organizing one of the first, I believe, joint gatherings of virologists, molecular biologists and enzymologists interested in the proteases of retroviruses. The possibility, first suggested by my long-time friend, Professor Jordan Tang of the Oklahoma Medical Research Foundation, was realized in the organization of Colloquium C 52, "Proteases of Retroviruses". We made every effort to invite to Prague the most experienced workers in this field and also to provide space for "latecomers" offering their own presentations. The interest in the colloquium during the first morning session took us by surprise. The original tight time schedule had to be expanded and a round-table discussion was organized the next morning. The immense interest in the colloquium confirmed us in the decision made before the Congress, that the presentations should not be documented by mere abstracts, a decision reinforced by the interest shown by several publishing houses in a volume containing the papers presented at the colloquium. Preference was given to Walter de Gruyter because of previous good contracts.
It is not up to me as editor to judge the quality of the book and I will leave judgment to the many readers I sincerely believe it will find, adding only one concluding comment. It is pointed out in Dr. Korant's lecture that researchers who regard viral proteases as "naive enzymes" should consider the possibility that they themselves are approaching these enzymes in a naive way. If this book helps the reader to obtain a less naive view of retroviral proteases I will feel that it has not been published in vain.

Prague, March 1989 Vladimír Kostka

Acknowledgments

The decision regarding the organization of the Colloquium was made at a relatively late stage in the preparation of the Congress. I wish to thank Prof. A. Kotyk, Chairman of the Scientific Program Committee, for the understanding with which he helped us to incorporate the Colloquium into the very tight Congress schedule. The organization of the Colloquium involved a great deal of work, for which my thanks go to Dr. P. Štrop. It is my pleasant duty to acknowledge gratefully the financial help of the following companies:

BioCarb Chemicals
Ciba Geigy
du Pont de Nemours & Co.
Merk, Sharp & Dohme
Pfizer
Roche Products Ltd.
Smith, French & Kline Laboratories

For this book I wish to thank all contributors, especially those (few) who submitted their manuscripts in time. It is my pleasure to express my thanks to Dr. J. Kostková, who prepared the indexes. I am very indebted to Ms. H. Pokorná for her typing work. Thanks are also due Mrs. E. Glowka of Walter de Gruyter, Berlin, for her understanding and cooperation.

Contents

Bacterial Expression, Processing and Characterization of Recombinant Retroviral Proteases

Mutational Analyses of Retroviral Proteases

Structure, Function and Evolution of Retroviral Proteases

CHARACTERISTICS OF VIRAL PROTEASES

S. I. Foundling and B. D. Korant

Central Research & Development Department, DuPont Company,
Experimental Station, Building 328, Wilmington, Delaware 19898, U.S.A.

Introduction

Viral proteases play an essential role in the replication process and also possess an extraordinary degree of substrate selectivity. These attributes make them appropriate targets for selective inhibitors. Such inhibitors could eventually be developed into antiviral agents and used clinically. The protease activity encoded by viruses was established with the picornaviruses (1) but, as will be described below, there are a number of important convergences in the properties of proteases for several animal virus groups, in particular the picornaviruses and retroviruses.

General Properties of Viral Proteases

It would be incorrect to suggest that all viral proteases display similar characteristics. However, for the picornaviruses and the retroviruses, there is an important set of properties held in common. First, the two classes of RNA viruses produce many of their proteins as polyproteins, i.e. in a large precursor format. The viral endo-protease which carries out the cleavages is itself initially a domain of the polyprotein, and is bounded by sites which it can process. One model for production of the enzyme is that it does so autocatalytically (2), and once released, proceeds to act as a protease at distant sites. This model is reasonable and fits most of the published data, but is by no means certain. The proposal (see below) that some retroviral proteases are active only as homodimers greatly complicates (but does not make inconceivable) the autocatalytic model.

The viral proteases tend to be significantly fewer in amino acids than their cellular counterparts (see (3) for a recent review), with that of HIV-1 containing fewer than one-half the residues of pepsin. This seems paradoxical in light of the atypically great selectivity of the viral enzymes for their substrates. These facts remain to be reconciled by one model.

Proteases of Retroviruses

There is no evidence to date that a viral protease has evolved a novel catalytic mechanism. The picornavirus protease is a member of the family (papain, cathepsin B) of cysteine active site proteases (4), while the retroviral protease is probably related to the aspartyl family, including pepsin and renin (5,6), but various individual reports cite sensitivity of retroviral proteases to inhibitors of other protease classes. Figure 1 summarizes the relevant similarities and some differences between the picornavirus and retrovirus protein processing systems.

Several models have been proposed for the folding unit of the retrovirus protease (Figure 2). The published model by Pearl and Taylor (7), constructed using pattern-recognition, template searching programs, structure prediction and molecular modelling techniques, concludes that the viral protease may adopt a fold resembling the single domain of an aspartic protease. The modelling is extended to infer that the protease would have to dimerise to conform to the structural restraints which are fundamental to the current ideas on catalysis by the aspartic proteases.

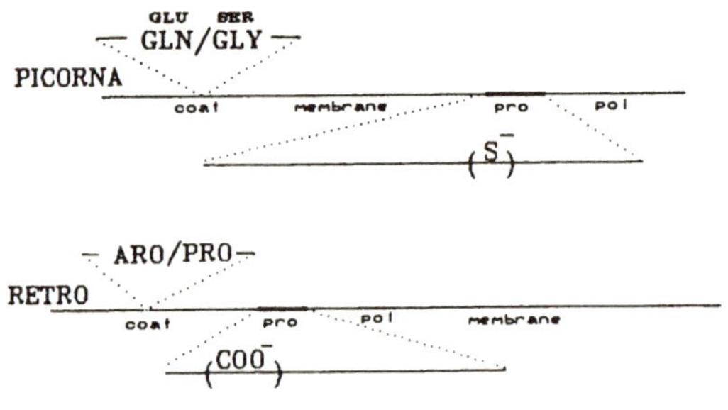

Figure 1. Genome organization and protease specificities of picornaviruses and retroviruses.

An alternate approach, taken by Blundell and co-workers (8), was to construct a generalized framework, modelling from the NH2- and COOH- terminal domains independently. A composite model was then generated, taking regions of greater homology from either domain framework and then merging loops of the appropriate length onto the averaged core of beta-strands, thus deriving the conserved tertiary folding unit of the viral protease.

The mode of substrate binding may share similarities to those observed for the aspartic proteases (9,10), but in the absence of equivalent residues to Thr/Ser 219 and Tyr 75 of the 'flap', there will be changes in the mode of substrate and inhibitor binding, which cannot be inferred from molecular models. The assumed 'dimer' of the viral protease (7), although providing a binding cleft of sufficient dimensions to accommodate four to six residues of substrate would, due to its 2-fold symmetry, possess an equality in binding specificity not only at the primary specificity sites, but also along the length of the cleft at the secondary binding sites.

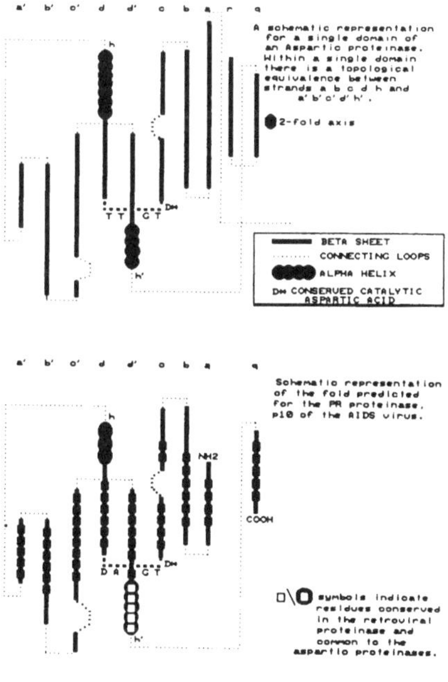

Figure 2. Model of the retrovirus pol protease.

It is difficult to envisage how such a dimer would function showing such exacting specificity for its polyprotein substrates. The aspartic proteases offer quite an asymmetric binding environment by comparison and effectively distort the substrate to aid in the catalytic event (11,12). Site-directed mutagenesis offers a direct way to test the hypothesis that the pol protease may function as a prototypic domain of an aspartic protease. Key residues to mutate would be the highly conserved or invariant ones. The sequence LLDTGAD in HIV-1 pol protease offers likely candidates, and Mous et al (13) have reported that a single amino acid change in LLDTGAD to LLATGAD leads to complete inactivation of the processing of the pol polyprotein, suggesting a fundamental role for this residue in the catalytic mechanism. It is possible that the pol protease and other retroviral proteases belong to a special category of proteolytic enzymes, but with a catalytic mechanism in common with the aspartic proteases.

Role of the Substrate in Cleavage of Viral Polyproteins

The cleavage sites of viral polyproteins are well-known in terms of their primary cleavage sites (P1 and P1' residues), which are glutamine-glycine for picornaviruses and hydrophobic-proline for retroviruses. However, there is little additional information on consensus sequences or structures recognized by viral proteases, and the limits of recognition, including local constraints as well as tertiary ones, need much additional study. It is clear that the viral processing systems are more sophisticated, or at least more selective, than the protein degrading enzymes, and perhaps are more analogous to the cellular proteases involved in complex regulatory events such as prohormone processing.

A common question regarding the viral proteases is whether they are able to act in trans, or if they must be a covalently-bound part of the polyprotein being processed. The answer is unequivocally that they can act in trans, if an appropriate substrate is provided to them, but they do so rather inefficiently. It is worth recalling that a typical viral protease is required to cleave only a few times (less than ten) in completing its processing function, so that high turnover rates are not selected for, but high specificity is.

From studies done with recombinant systems (14,15) and synthetic peptide substrates, as well as inspection of the processing sites published in many individual papers, the sequence and suggested structural model of a typical picornavirus cleavage site is shown in Figure 3.

The ability to clone a viral protease and its substrate has led us to conclude that while the picornavirus protease may bind in trans, there is probably little basis for the interaction built into the tertiary structure of the polyprotein. This is because the polyprotein can be rearranged, scrambled, or have large portions deleted, and still be processed at authentic sites (15).

In part because of those results, we propose a model for the action of the viral protease consisting of a translocating/scanning mode (see Figure 4), in which the enzyme, after autocatalytic excision, remains bound to the polyprotein and travels along its length, cutting where it finds a homologous cleavage sequence of eight to ten amino acids. At the present time, there is little direct evidence for this model. Also to be addressed is the problem of how energy would be made available to drive the translocation.

Through the action of the retrovirus protease on pol, functional units are liberated; the reverse transcriptase p51 (RT), exo-nuclease-ribonuclease H p15 (RNase H) and endonuclease-integrase p32 (IN). There is limited sequence information concerning the cleavage between RT and RNase H. Similarly, the junction for the cleavage between RNase H and IN is also unknown, but if other cleavage recognition sequences are used to bias a prediction for these clip sites within pol and gag, then a conservative sequence specificity starts to emerge for the protease.

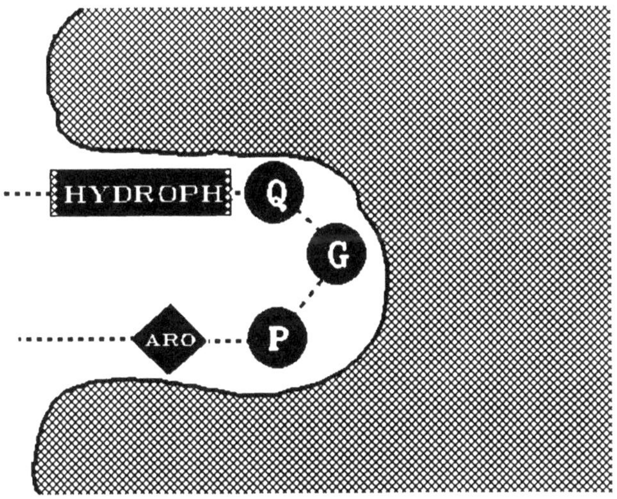

Figure 3. Model of the structure of the picornavirus polyprotein cleavage site. Cross-hatched area represents the viral protease.

Table 1. Known and predicted cleavage sites for pol and gag polyproteins catalyzed by the pol protease. Cleavages occur between P1 and P1' residues.

Pol Proteins			P3	P2	P1	P1'	P2'	P3'	
p5:p10PR	V	S	F	N	F	P	Q	I	T
PRp10:p51 RT	C	T	L	N	F	P	I	S	P
RTp51:p15 RNaseH?	Q	L	E	K	E	P	I	V	G
RNaseHp15:p32 IN?	I	R	I	I	L	F	L	D	G
Gag Proteins									
MA p17:p25 CA	V	S	Q	N	Y	P	I	V	G
CA p25:p15 NC?	K	A	R	V	L	A	E	A	
CA p25:p15 NC	T	A	T	I	M	M	Q	R	G

Two cleavage sites are indicated for CA p25:p15 NC: the COOH- terminal sequence of p25 CA and NH2- terminus of p15 NC are 14 residues apart. Nomenclature: PR, proteinase; RT, reverse transcriptase; RNase H, exonuclease-ribonuclease H; IN, endonuclease-integrase; MA, matrix protein; CA, capsid protein; NC, nucleocapsid protein (16).

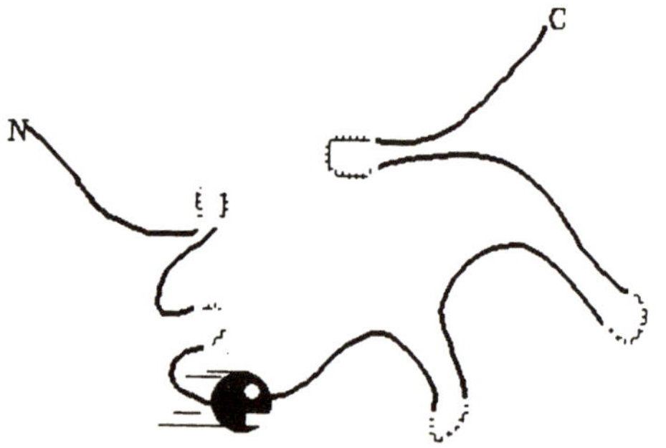

Figure 4. Scanning/translocating model for the action of the picornavirus protease.

The residues adjacent to the bond cleaved, i.e. the primary specificity residues, are predominantly hydrophobic with a preference for Pro at P1'. The only exception to the striking specificity towards hydrophobic residues is at the predicted junction between p51 RT and p15 RNaseH, where there is a Glu substitution at P1. A consensus is also observed for residues at P3 to P2 and P2' to P3', the secondary specificity residues. The P2 position shows preference for aliphatic/small polar groups and P2' for hydrophobic/large polar groups. The specificities of residues at P3 and P3', however, fall into no distinct category. The sequence specificity is not unlike that observed for the aspartic proteases (12). These enzymes show a preference for cleaving between hydrophobic residues, but the binding cleft has evolved to accommodate up to nine residues of substrate. Some members of the group, e.g. human renin and calf chymosin, show high specificity for the residue type accommodated at each binding pocket in the cleft, reflecting the high specificity shown for their physiological substrates.

Strategies for Inhibitor Design

It is becoming obvious that viral proteases offer targets for specific inhibitors, and possibly for antiviral drugs (17). A current list of various approaches to antivirals based on protein cleavage inhibitors is given in Table 2.

Table 2. Classes of Viral Protein Cleavage Inhibitors

Class	Antiviral Activity
1. Substrate Modifiers	
Amino acid analogs	Millimolar
Phenethyl alcohol	Millimolar
Metal ions (zinc)	Micromolar
Chelating agents	Millimolar
High temperature(>37°)	–
2. Endogenous Inhibitors	
Cystatins	Micromolar
3. Synthetic Inhibitors	
Pepstatin	Micromolar
Transition state analogs	Micronanomolar
Epoxides	Micromolar
Cleavage site mimics	Micromolar

Since the viral proteases are apparently mechanistically in classes already occupied by important cellular enzymes, selectivity will have to be built into the compounds based on the substrate specificity of viral proteases we have discussed in previous sections of this report. One approach is to design peptidic compounds which contain viral cleavage site sequences and appropriate targetting substituents, as well as groups which react with catalytic residues of the protease. Such an approach has already been described (18). The problem of pharmaco-kinetics, in particular stabilization of such peptides in animals, must be addressed if the peptides are to become clinically useful drugs.

Blocking viral processing by inhibiting the retroviral protease may lead to non-infective, immature virions (19). The development of antivirals is likely to reflect the successful development of inhibitors for other aspartic proteases, for example, antihypertensives directed against human renin. Given that pepstatin A, a naturally occurring acylated pentapeptide inhibitor of the aspartic proteases, shows inhibitory activity against the viral protease, the immediate avenue to explore will place the unusual amino acid, statine, into the natural cleavage sequence for the viral protease. Using this approach, tight binding inhibitors were developed for human renin (20), in which the amino acid residues at positions P1 and P1' of the cleavage site in human angiotensinogen were replaced by the statine dipeptide analog. An alternative approach may be the incorporation of the statine mainchain chemistry with modified cyclohexyl sidechain to mimic the Phe-Pro bond cleavage between PR p10 and p51 RT.

Using the synthetic approach to inhibitor design through the modification of the scissile bond between residues P1 and P1' with the incorporation of non-hydrolyzable transition state analogs of this bond, highly potent compounds may be developed. Inhibitors designed with these principles are excellent inhibitors of human renin (21).

Another potentially fruitful area for study is the role of endogenous inhibitors, such as the well-known cystatins, which inhibit cysteine proteases (22). Analogous inhibitors for aspartic proteases are not yet available from human tissues. The possible role of such inhibitors in intrinsic resistance to viral infections is of increasing interest. We have previously suggested a protocol in which a purified protease and antibodies to the enzyme may be used to screen expression libraries for protease ligands (and by inference, inhibitors). The procedure has been shown to be workable for a model system in which purified poliovirus protease was used as a probe for chicken cystatin and human cystatin C (22).

Conclusions

The history of chemotherapy for viral infections is not filled with great successes. Except for amantadine (influenza A) and acycloguanosine (herpes), the few other approved antiviral drugs are used rarely because of unacceptable side effects or low efficacy. Azidothymidine (AZT) is an example of an antiviral with a marginal therapeutic ratio, but it is the only drug approved for AIDS patients presently, and its use is on the increase.

AIDS is at the same time a fearsome disease and an opportunity for molecular biology to contribute to human health. The enzymes and regulatory proteins of the retroviruses show much greater genetic conservation than the viral structural antigens, and offer targets which are approachable in terms of chemical design. The challenge in the coming years will be to choose one of the viral enzymes and try to develop an inhibitor which is selective, well-tolerated, and effective, in the face of a very complicated, persistent infection, which progressively disintegrates the host's defense mechanisms and enters and departs from poorly-defined latencies in a battery of tissues, including the nervous system. In light of such complexity, the successful development of new antiviral drugs will be difficult, but AZT has shown that it is possible to do so. The need and potential benefits are great, in particular if the number of AIDS cases in the developing countries and in the heterosexual populations of developed countries continue to increase significantly. It may also be anticipated that (retro) viral diseases will continue to be discovered, as new probes are applied to old and emerging diseases, and they may provide additional needs for chemotherapeutic intervention.

References

1. Korant, B. 1972. J. Virol. 10, 751.

2. Hanecak, R., B. Semler, H. Ariga, C. Anderson, E. Wimmer. 1984. Cell 37, 1063.

3. Korant, B. J. Antibiotics (in press)

4. Ivanoff, L., T. Towatari, J. Ray, B. Korant, S. Petteway. 1986. Proc. Nat. Acad. Sci. U.S.A. 83, 5392.

5. Von der Helm, K. 1977. Proc. Nat. Acad. Sci. U.S.A. 74, 911.

6. Yoshinaka, Y., I. Katoh, T. Copeland, S. Orozlan. 1985. Proc. Nat. Acad. Sci. U.S.A. 82, 1618.

7. Pearl, L., W. Taylor. 1987. Nature 329, 351.

8. Blundell, T., D. Carney, S. Gardner, F. Hayes, B. Howlin, T. Hubbard, J. Overington, D. Singh, B. Sibanda, M. Sutcliffe. 1988. Europ. J. Biochem. 172, 513.

9. James, M., A. Sielecki. 1985. Biochemistry 24, 3701.

10. Blundell, T., J. Cooper, S. Foundling, D. Jones, B. Atrash, M. Szelke. 1986. Biochemistry 26, 5585.

11. Pearl, L. 1987. FEBS Lett. 214, 8.

12. Foundling, S. 1987. Doctoral Thesis, Univ. of London.

13. Mous, J., E. Heimer, F. LaGrice. 1988. J. Virol. 62, 1433.

14. Ypma-Wong, M., B. Semler. 1987. Nucl. Acids Res. 15, 2069.

15. Cordova, A., B. Korant In: Intracell. Protein Catabol. 7 (N. Katunuma, ed.) Japan Sci. Soc. Press, Tokyo (in press).

16. Leis, J., D. Baltimore, J. Bishop, E. Fleissner, S. Goff, S. Orozlan, H. Robinson, A. Skalka, H. Temin, V. Vogt. 1988. J. Virol. 62, 1808.

17. Korant, B. 1981. In: Antiviral Chemotherapy: Design of Inhibitors of Viral Functions (K. Gauri, ed.) Academic Press, N.Y. p. 37.

18. Kettner, C., B. Korant. 1987. U.S. Patent 4,644,055.

19. Katoh, I., T. Yasunga, Y. Ikawa, Y. Yoshinaka. 1987. Nature 329, 654.

20. Boger, J. 1985. In: Aspartic Proteinases and Their Inhibitors (V. Kostka, ed.) De Gruyter, Berlin, p. 401.

21. Szelke, M., B. Leckie, A. Hallett, D. Jones, J. Suieras, B. Atrash, A. Lever. 1982. Nature 299, 555.

22. Korant, B., T. Towatari, M. Kelley, B. Lenarcic, J. Brzin, V. Turk. 1988. Biological Chem. H.-S. 369 (suppl.), 281.

MORPHOGENESIS OF RETROVIRUSES IN THE PRESENCE AND ABSENCE OF PROTEASE INHIBITORS

R.B. Luftig, M. Bu, K. Ikuta

Department of Microbiology, Immunology and Parasitology, Louisiana State University Medical Center, New Orleans, Louisiana 70112-1393

Introduction

Retroviral proteases from both murine leukemia virus (MuLV) and human immunodeficiency virus (HIV) are viral coded enzymes that cleave *gag* and *gag-pol* precursor polyproteins into functional products, needed for assembly of mature virus particles. An expression vector obtained from Dr. Ron Swanstrom (University of North Carolina) containing an open reading frame which starts at the 5' end of the *pol* gene was used to provide an 11 kd product that specifically cleaved Gazdar (Gz) MuLV Pr65^{gag} into Pr40^{gag} (p30+p10), Pr27^{gag} (p15+p12) intermediates, as well as the smaller Mr products, such as p30. Using an immunoblotting system with MuLV anti-p30 or p12 sera; pepstatin A, cerulenin and some analogues of cerulenin were observed to inhibit HIV protease cleavage by greater than 50% at concentrations ranging from (0.2-0.5 mM).

Results

Over the past few years our laboratory has found that cerulenin, an anti-fungal antibiotic, inhibits MuLV replication because it prevents not only Pr65^{gag}, but the *env* precursor Pr80^{env}, as well, from being cleaved (1). Cerulenin itself is a small (MW:223) antifungal antibiotic produced by the fungus *Cephalosporium caerulens* that was originally shown to inhibit *de novo* fatty acid and cholesterol biosynthesis (2). Goldfine, *et al.* (3) had originally observed that cerulenin inhibited Rous sarcoma virus (RSV) production while failing to process the *gag* RSV Pr76^{gag}. Thus, cerulenin

Proteases of Retroviruses

appears to have a two-fold inhibitory mode of action on MuLV assembly; a) by decreasing the pool of fatty acids, such as myristic acids, which are required to be covalently attached to the NH_2 terminus of MuLV $Pr65^{gag}$ (4), and b) acting as a protease inhibitor. We have shown in support of these expected results that, when 20 µg/ml cerulenin is added to MuLV infected cells, there is a 2-3-fold accumulation of immature budding viruses containing an excess of uncleaved $Pr65^{gag}$ (5). Also, when $Pr65^{gag}$ is present in an aberrant or mutant form, such as in 3JE cells; then, after cerulenin treatment, the unmyristylated $Pr65^{gag}$ is rapidly degraded (6). These results have led us to examine the potential anti-protease activity of cerulenin, *in vitro*.

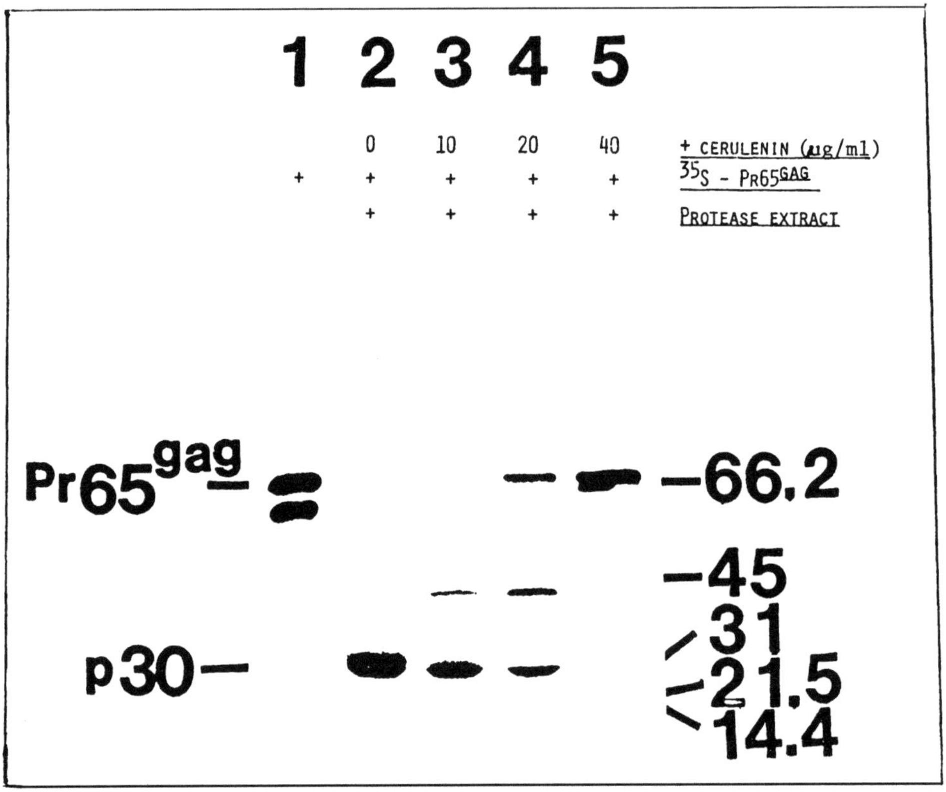

FIGURE 1

As can be seen above, in Figure 1, using ^{35}S-methonine-labeled $Pr65^{gag}$ from Gz-MuLV particles as a substrate, cerulenin at 20-40 µg/ml (corresponding to 0.1-0.2 mM) inhibits cleavage of Pr65^{gag} by added protease. As we go from lanes 2-5 in Figure 1, we note at increasing concentrations of cerulenin that Pr65^{gag} cleavage to p30 and a 40kd intermediate is increasingly blocked. From computer analyses, using the GENBANK sequence base, we know that the HIV and MuLV *gag* precursor cleavage sites are the same, i.e., (tyr,pro), (phe,pro). Also, the HIV-coded *gag* protease shares >60% homology with the MuLV Pr65^{gag} protease (including its putative active site and a potential dimer forming sequence). Thus, we thought that cerulenin or its analogs would be useful drugs to employ as inhibitors of the HIV protease. Toward this goal, we have established an *in vitro* assay system that utilizes MuLV Pr65^{gag} from Gazdar virus as a substrate and bacterially produced HIV protease (DNA plasmid expressing most of the *pol* open reading frame obtained from Dr. Swanstrom) as the enzyme. We first confirmed the presence of HIV RT in IPTG induced extracts using AIDS patient sera. Further, using an antisera made against a peptide at the 3' end of the protease gene product, we also identified an 11-12 kd protein by Western Blotting. Additionally, IPTG induced bacterial extracts were shown to contain an activity that specifically cleaved the Moloney MSV (Gazdar) Pr65^{gag} precursor into major 27 kd, as well as other lower MW *gag* products. This cleavage was blocked by 0.5 mM pepstatin, 0.2 mM cerulenin and $<$0.1 mM conentrations of two cerulenin analogs. Currently, we are examining the *in vitro* effect of these inhibitors on MuLV as well as HIV replication.

Acknowledgement

This research was supported in part by Public Health Service Grant CA-37380 from the National Cancer Institute and AMFAR Grant No. 000739 D. We thank Dr. Ron Swanstrom for his gift of the HIV plasmids.

References

(1) Ikuta, K., R.B. Luftig: 1986. Virology 154, 195.

(2) Omura, S.: 1976. Bacteriol. Rev. 40, 681.

(3) Goldfine, H., J.B. Harley, J.A. Wyke: 1978. Biochem. Biophys. Acta 512, 229.

(4) Rein, A., M.R. McClure, N.R. Rice, R.B. Luftig, A.M. Schultz: 1986. Proc. Natl. Acad. Sci. USA 83, 7246.

(5) Katoh, Y., Y. Yoshinaka, R.B. Luftig: 1986. Virus Research 5, 265.

(6) Ikuta, K., J. Coward, R.B. Luftig: 1986. Virology 154. 207.

PROCESSING OF THE ENVELOPE AND CORE PROTEINS OF HUMAN IMMUNODEFICIENCY VIRUS-TYPE-1

M.G. Sarngadharan, Ranajit Pal

Department of Cell Biology, Bionetics Research, Inc., Rockville, Maryland 20852

Robert C. Gallo

Laboratory of Tumor Cell Biology, National Cancer Institute,
Bethesda, Maryland 20892

Introduction

The gag and env gene products of the human immunodeficiency virus type 1 (HIV-1) are synthesized in the form of precursor polyproteins Pr^{53gag} and gp160 respectively. Pr^{53gag} undergoes post translational cleavage to structural proteins of molecular weights 17,000, 24,000 and 15,000. Gp160 is cleaved into gp120, the amino terminal exterior component and gp41, the carboxy terminal transmembrane portion (1). The major glycosylation pathway utilized by the viral and cell membrane glycoproteins begins with the transfer of the precursor oligosaccharide Glc_3-Man_9-$GlcNAc_2$ from the carrier lipid, dolichol phosphate to the nascent polypeptide chain of the glycoprotein (2). The complex type oligosaccharides are generated via the trimming pathway. The first step of such processing is the removal of three glucose residues by glucosidases I and II. Subsequently the action of mannosidase I in the Golgi complex completes the removal of four mannose residues by mannosidase II and the addition of peripheral sugar

Proteases of Retroviruses

residues like galactose, N-acetyl glucosamine and sialic acid by specific glycosyl transferases result in the synthesis of complex type oligosaccharide moieties in the glycoprotein molecule (1). Several inhibitors of the trimming enzymes are now available (3). Thus 1-deoxynojirimycin and castanospermine have been shown to inhibit glucosidase I activity in the endoplasmic reticulum and 1-deoxymannojirimycin and swainsonine inhibit mannosidases I and II respectively in the Golgi complex. We studied the effects of these compounds on the processing and the cleavage of gp160 in HIV-infected cells. We also report the effects of monensin an agent that is known to affect the transport and expression of membrane glycoproteins (4) and the antibiotic cerulenin, an inhibitor of de novo fatty acid and sterol synthesis (5) on the processing of HIV-1 envelope and core proteins.

Results

Effects of tunicamycin, deoxynojirimycin, deoxymannojirimycin and swainsonine on the synthesis and processing of gp160.

The effect of specific inhibitors of oligosaccharide processing on the synthesis of viral proteins in HIV-1-infected cells was studied using Molt3/HTLV-III_B cells. The cells were labeled with [^{35}S]-methionine in the presence of the inhibitors, and the immunoreactive proteins in the lysates were precipitated with a HIV-1 antibody-positive human serum. As depicted in Fig. 1, 1.5 μg/ml of tunicamycin (Calbiochem) severely inhibited the glycosylation of the envelope protein resulting in the accumulation of an 80 kilodalton molecule (lane 4). Treatment of cells with 5 mM deoxymannojirimycin (MNM) (Boehringer Mannheim) or 10 μg/ml of swainsonine

(Boehringer Mannheim) had no effect on the processing of gp160, although a major 100 kDa protein was observed in

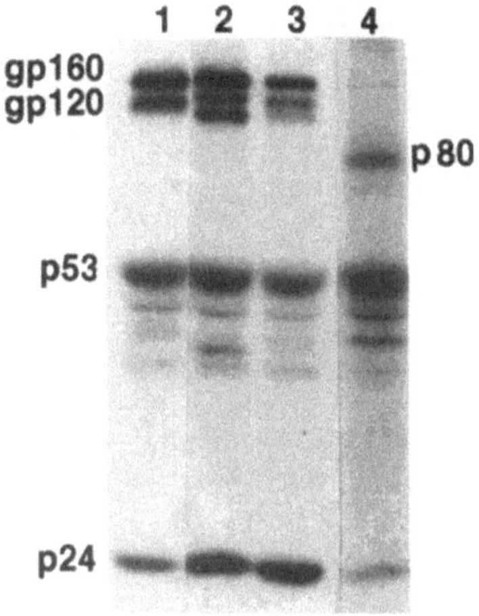

Fig. 1. The profile of viral proteins from HIV-1-infected cells in the presence of deoxymannojirimycin, swainsonine, or tunicamycin.

each case in addition to the usual gp120 (Fig. 1, lanes 2 and 3). On the other hand, 5 mM deoxynojirimycin (DNM) inhibited the proteolytic processing of gp160. In a pulse-chase experiment the cells were labeled with [^{35}S] methionine in the presence of 5 mM DNM for 1 hr, and were subsequently chased for up to 4 hr by the addition of cold methionine (0.1%) in the same medium. As shown in Fig. 2, there was a significant processing of both env and gag precursor proteins in untreated control cells (lanes 1 to 3). However, cells radiolabeled in the presence of DNM had a large proportion of gp160 which could not be chased to gp120 even after 4 hr (Fig. 2, lanes 4 to 6). None of the inhibitors studied affected the processing of the gag precursor Pr^{53gag} to p24.

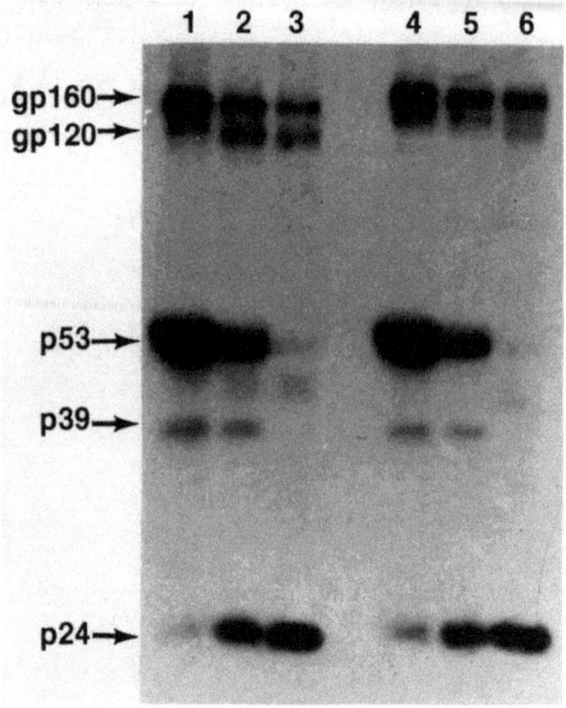

Fig. 2. Analysis of the metabolic fate of HIV-1 glycoproteins in Molt3/HTLV-III_B cells treated with deoxynojirimycin

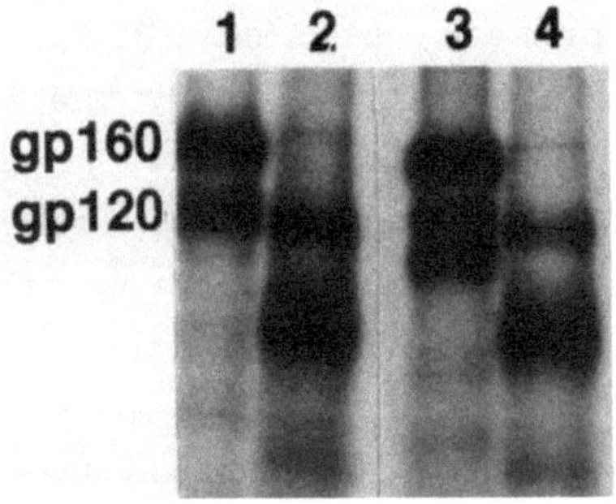

Fig. 3. Susceptibility of glycoproteins from HIV-1-infected cells to endoglycosidase H digestion. Lanes 1 and 2 represent control cells and 3 and 4 MNM-treated cells. Immunoprecipitates in lanes 2 and 4 were endo H treated.

Cleavage of gp160 occurs before the trimming of mannose residues.

As described earlier, the biosynthesis of an N-glycosylated protein starts with the addition of an oligosaccharide to asparagine residues and is followed by stepwise trimming of glucose and mannose residues and the addition of peripheral sugars. At what stage of sugar trimming does HIV gp160 undergo cleavage to gp120 and gp41? This question was analyzed by examining the sensitivity of the three HIV-1 glycoproteins to endoglycosidase H (Endo H) treatment. Endo H specifically cleaves N-linked oligosaccharides containing cores with high mannose content, such as is present before the mannose trimming. In contrast, the complex oligosaccharides formed in the Golgi are not affected by Endo H (2). Molt3/HTLV-III_B cultures were labeled with [^{35}S]-methionine in the presence and absence of MNM and the labeled cells were extracted with detergents. The HIV-1 proteins in the lysates were immunoprecipitated with a broadly reactive serum from a HIV-1-infected subject. Portions of the immune precipitates were digested with endo H and then analyzed by SDS-polyacrylamide gel electrophoresis along with their undigested controls. The results are described in Fig. 3. Irrespective of whether the culture was treated with MNM or not, gp160 was always totally sensitive to endo H treatment (Fig. 3), indicating that gp160 exists always in the high mannose state. On the other hand gp120 from untreated cells was only partially sensitive to endo H treatment. However, the 100 kDa gp120 species of the MNM treated culture contained mannose-rich cores as it was totally digested by Endo H. It is clear that the reduced size of this protein is the result of a block in the normal maturation of the sugar moieties to the complex carbohydrate type. Both gp120 and the 100 kDa variant gave rise to the same size product after endo H treatment.

The carbohydrate residues of gp41 exist mostly in the complex state and therefore were largely resistant to endo H digestion (data not shown). However, as expected, gp41 from cells treated with MNM contained mannose-rich sugar residues as evidenced by the total sensitivity to endo H digestion.

Synthesis of HIV-1 proteins in Molt3/HTLV-III_B cells treated with monensin.

Molt3/HTLV-III_B cells treated with 7.5 μM monensin were radiolabeled with $[^{35}S]$-methionine for 1 hr and the label

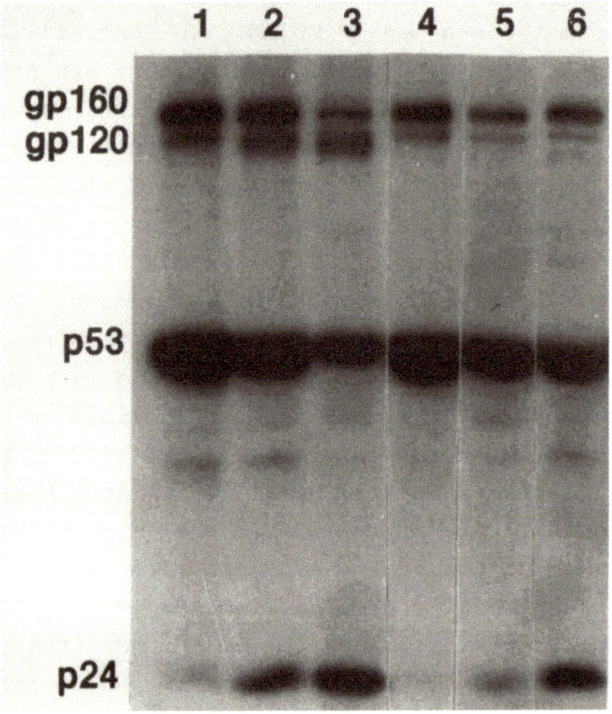

Fig. 4. Analysis of the metabolic fate of HIV-1 proteins in untreated and monensin-treated cells. The chase times were 0 (lanes 1 and 4), 60 (lanes 2 and 5) and 240 min (lanes 3 and 6)

was subsequently chased up to 4 hr. The cells were harvested at various times and the level of intracellular viral proteins was determined by immunoprecipitation with a human serum. In untreated control cells there was a significant decrease in both gp160 and Pr^{53gag} with corresponding increases in gp120 and p24 respectively during the chase (Fig. 4, lanes 1-3). In sharp contrast, in monensin-treated cells only the gag proteins underwent normal processing. There was a marked reduction of the proteolytic cleavage of gp160 (lanes 4-6). The labeled gp160 that accumulated in monensin-treated cells could not be processed in the presence of a lysate of untreated infected cells suggesting that the processing of gp160 took place in intact organelle and could not be accomplished in cell-free extracts (6).

Synthesis of HIV-1 proteins in Molt3/HTLV-III_B cells treated with cerulenin.

To determine whether cerulenin affects the processing of HIV-1 proteins, Molt3/HTLV-III_B cells were labeled with [^{35}S]-methionine in the presence of different concentrations of cerulenin (CalBiochem). The labeled cells were harvested and their extract immunoprecipitated with an HIV-1 antibody positive human serum. The result of such an immunoprecipitation reaction is depicted in Fig. 4. There is a general inhibition of viral protein synthesis at higher concentrations of cerulenin (Fig. 5, Panel A). However, the proteolytic cleavage of Pr^{53gag} to p24 was selectively inhibited in cerulenin treated cells. Interestingly, at a concentration of cerulenin that virtually stopped the cleavage of p53 to p24 (10 μg/ml), myristylation of the gag precursor was still detected, albeit at a low level (Fig. 5, Panel B). The stability of the gag polyprotein p53 accumulated following cerulenin

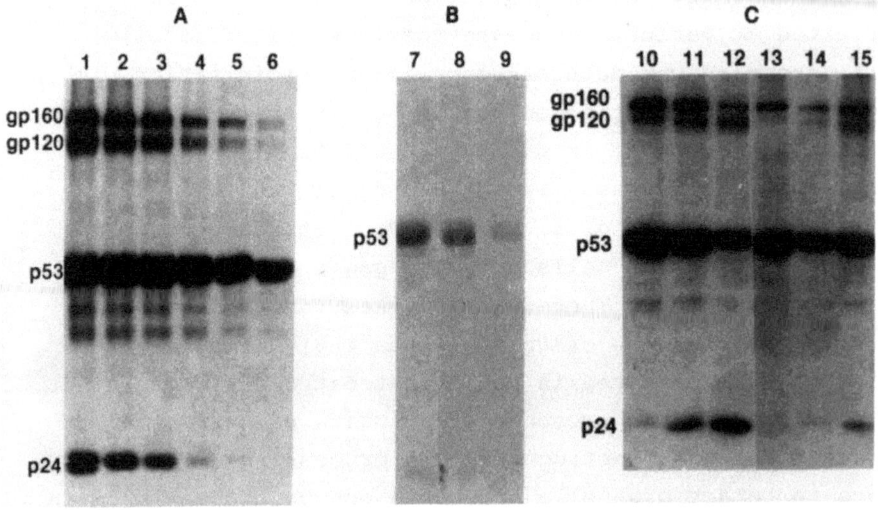

Fig. 5. Intracellular protein profile of Molt3/HTLV-III$_B$ cells treated with various concentrations of cerulenin. Panel A. Concentrations of cerulenin (μg/ml) in lanes 1-6 were: 0, 2.5, 5, 7.5, 10 and 12.5. Panel B. Molt3/HTLV-III$_B$ cells were labeled with [^{3}H]-myristic acid (100 μCi/ml) for 7 hr. Cerulenin was absent (lane 7) or present at 7.5 (lane 8) or 10 μl/ml (lane 9). Panel C. Analysis of metabolic fate of HIV-1 proteins in the presence of cerulenin. Chase times were 0 (lanes 10 and 13), 60 (lanes 11 and 14) and 240 min (lanes 12 and 15).

treatment was determined by pulse-chase experiments. Molt3/HTLV-III$_B$ cells were radiolabeled for 1 hr in the presence of 7.5 μg/ml of cerulenin and the label was then chased for up to 4 hr. The cells were harvested at various times and the intracellular protein profile determined by immunoprecipitation. As shown in Fig. 5, Panel C, Pr^{53gag} was processed to p24 in untreated Molt3/HTLV-III$_B$ cells (Lanes 10-12). However, no significant processing of p53 was observed under similar conditions in cerulenin treated cells (Lanes 13-15).

Discussion

The results presented here demonstrate that in HIV-1-infected cells the cleavage of gp160 to gp120 takes place in proteins containing unprocessed oligosaccharide chains. Thus the uncleaved gp160 is totally sensitive to endo H digestion and is in a high mannose state irrespective of the presence of mannosidase inhibitors in the culture. Such a total endo H sensitivity of gp120 and gp41 was only detected in cells treated with MNM. The oligosaccharide structures linked to gp120 have recently been shown to carry oligomannosidic glycans with eight or more mannose residues (7). The cleavage of a precursor glycorpotein containing mannose-rich oligosaccharide structures was also noted in other enveloped viruses (8).

We have demonstrated that in Molt3 cells chronically infected with $HTLV\text{-}III_B$, monensin inhibited the proteolytic processing of gp160 to gp120. However, the limited processing that occurs in these cells lead to a 100 kDa product in addition to gp120 as was seen in the present study with MNM treated cells. This suggests that the cleavage of gp160 takes place at an intracellular site, probably within the Golgi complex.

In this communication we have demonstrated that the treatment of $HTLV\text{-}III_B$-infected Molt3 cells with cerulenin inhibited the cleavage of Pr^{53gag} to p24. Myristylation of Pr^{53gag} was also inhibited by cerulenin. It is not clear how the cleavage of p53 was inhibited by cerulenin. One possible explanation is that the antibiotic acts as an inhibitor of viral protease which is responsible for processing the *gag* polyprotein. Another possibility is that the myristylation of the *gag* protein is essential for its interaction with intracellular membranes where the cleavage of the p53 to p24 takes place. Inhibition of myristylation by cerulenin might affect the interaction of the gag polyprotein with the membrane, thereby affecting its accessibility to the protease.

References

1. Sarngadharan, M.G., Markham, P.D. 1987. In: Acquired Immunodeficiency Syndrome and Other Manifestations of HIV-Infection (G.P. Wormser, R.E. Stahl, and E.J. Bottone, eds.). Noyes Publications, p. 186.

2. Kornfeld, R., Kornfeld, S. 1985. Ann.Rev.Biochem. 54, 631

3. Elbein, A.D. 1987. Methods Enzymol. 138, 661.

4. Tartakoff, A.M. 1983. Cell 32, 1026.

5. Schlesinger, M.J., Malfer, C. 1982. J.Biol.Chem. 257, 9887.

6. Pal, R., Gallo, R.C., Sarngadharan, M.G. 1988. Proc.Natl.Acad.Sci.U.S.A. 85, 9283.

7. Geyer, H., Holschbach, C., Hunsmann, G., Schneider, J. 1988. J.Biol.Chem. 263, 11760.

8. McDowell, W., Romero, P.A., Datema, R., Schwarz, R.T. 1987. Virology 161, 37.

MECHANISMS OF RIBOSOMAL FRAMESHIFTING FOR SYNTHESIS OF THE PROTEASE IN HIV AND OTHER RETROVIRUSES AND THE POSSIBLE USE OF HYPOMODIFIED tRNAs IN THE FRAMESHIFT EVENT

Dolph L. Hatfield, Byeong J. Lee

Laboratory of Experimental Carcinogenesis, National Cancer Institute, National Institutes of Health, Bethesda, MD 20892 USA

Ya-Xiong Feng, Judith G. Levin

Laboratory of Molecular Genetics, National Institute of Child Health and Human Development, National Institutes of Health, Bethesda, MD 20892 USA

Alan Rein, Stephen Oroszlan

Laboratory of Molecular Virology and Carcinogenesis, Bionetics Research, Inc., Basic Research Program, National Cancer Institute-Frederick Cancer Research Facility, Frederick, MD 21701

Introduction

The proteases of many vertebrate retroviruses are synthesized by a mechanism involving ribosomal frameshifting. The gag-pro frameshift site of MMTV (1) and the gag-pol frameshift site of HIV (2) have been studied in detail. Both studies show that the shift to the -1 reading frame occurs at a leucine codon. The signal for the frameshift event, however, appears to also involve the codons upstream of the frameshift site. Indeed, two conserved sequences, AAAAAAC and UUUA, which occur in the overlapping reading frames of numerous retroviruses, have been proposed as signals (3). To elucidate the frameshift sites and signals in expression of the protease and polymerase and of the tRNAs utilized for translating codons within these sites and signals, we searched the overlapping reading frames of vertebrate retroviruses for potential frameshift signals and then analyzed the chromatographic properties of aminoacyl-tRNAs which may be involved in the frameshift event.

Proteases of Retroviruses

Results and Discussion

We examined the overlapping reading frames of vertebrate retroviruses for potential ribosomal frameshift sites using the two conserved sequences, AAAAAAC and UUUA, which have been proposed as signals for the frameshift event (3). The gag-pro overlapping reading frames of MMTV (1,3,4), BLV (5,6), HTLV-1 (7) and HTLV-2 (8) contain a homopolymeric A region, AAAAAAC, while the gag-pol overlaps of HIV-1 (9,10), HIV-2 (11), STLV-III (12), SIV_{MAC} (13), and RSV (14,15) and the pro-pol overlaps of MMTV (4), BLV (5, 6) HTLV-1 (7) and STLV-I (7) contain UUUA (see also 3). The two conserved sequences and the codons surrounding these sequences within the overlapping reading frames of several retroviruses are shown in figure 1. We aligned the sequences vertically to facilitate the search for commom properties.

VIRUS	ORF	0 FRAME				-1 Frame
		9 8 7	6 5 4	3 2 1	a b	c d e
MMTV	gag-pro	U C A	A A A	A A C	U U	G U A
HTLV-1	gag-pro	C C A	A A A	A A C	U C	C A U
BLV	gag-pro	U C A	A A A	A A C	U A	A U A
HIV-1	gag-pol		A A U	U U U	U U	A G G
HTLV-1, BLV	pro-pol		U U C	C C U	U U	A A A
MMTV	pro-pol		C A G	G A U	U U	A U G
RSV	gag-pol		A C A	A A U	U U	A U A

Fig. 1. Ribosomal frameshift sites and signals in several retroviruses (see text for references). ORF designates overlapping reading frame, underlined regions designate conserved sequences within these overlaps (see text), 0 FRAME designates codons read within the gag frame and -1 FRAME designates codons read after the shift to the -1 reading frame.

The location and boundaries of the frameshift sites in MMTV and HIV-1 have been established by sequencing the transframe peptide generated in vivo from the gag-pro overlapping region in MMTV (1) and the transframe peptide generated in vitro from the gag-pol overlapping region in HIV-1 (2). In MMTV, the frameshift occurs at a leucine codon (UUG) which is immediately downstream of the AAAAAAC sequence (1). The asparagine codon (AAC) which is coded by positions 3, 2 and 1 in the figure is read in the zero frame. The 3' G in UUG (which is base c in the figure) and the next two downstream bases (U and A; bases d and e in the figure) code for valine which is read in the -1 frame (1). The occurrence of a leucine codon (UUG) at the frameshift site in MMTV, a serine codon (UCC) in HTLV-1 and a termination codon (UAA) in BLV (see bases designated a, b and c in the figure) suggests that the base in position c is not part of the frameshift site. That is, it seems unlikely that either: 1) the tRNA which causes the frameshift could decode UUG, UCC and UAA; or 2) a large number of tRNAs (leucine, serine and an ochre suppressor tRNA) are capable of carrying out the frameshift event. Thus, the frameshift site in retroviruses utilizing the homopolymeric A sequence as a signal probably does not involve the base in position c.

In HIV, the UUUA sequence is involved in the frameshift (see fig. 1) and the shift occurs at a leucine residue corresponding to the UUA codon (2). Both leucine and phenylalanine occur at the frameshift site in a 7:3 ratio which makes the assignment of the precise site uncertain (2). The phenylalanine codon (UUU) which occurs upstream of the frameshift site in HIV (and is coded by positions 3, 2 and 1 in the figure) is read in the zero frame (2). Arginine is coded by the 3' base of the UUA sequence and the first two GG bases in the downstream codon and arginine is translated in the -1 frame (see bases c, d and e in fig. 1) (2). Since the base designated c in the figure is decoded in the -1 frame, the frameshift probably occurs immediately upstream of this position. The report that mutations in position c do not inhibit frameshifting in RSV (P.M. Pryciak, T.E. Jacks and H.E. Varmus, personal communication) supports this conclusion. Thus, the frameshift site in retroviruses utilizing UUUA as a signal, like that in cases using AAAAAAC, probably does not include the base in position c.

Since the frameshift event occurs after decoding of the triplet at positions 3, 2 and 1 and does not appear to involve base c, the frameshift apparently takes place at bases a and b or possibly at bases 1, a and b (see below). The base in position a is a U in each frameshift site (see fig. 1) and this base appears to be invariant. The base in position b is variable. However, since the base in position c is not part of the frameshift site, then the base designated b must be decoded by the wobble position of the tRNA anticodon. Decoding of the frameshift site to achieve the -1 reading frame where the latter base represents the 3' end of the codon may occur in one of two ways. One possibility is a two out of three base translation in which only the bases designated a and b are read. One attractive feature of proposing doublet base decoding is that such a mechanism of shifting the reading frame in the -1 direction occurs in bacteria (16); and hence, there is a precedent for such a proposal. The procaryotic mechanism involves decoding at GCA (alanine) and CCG (proline) codons by serine and threonine tRNAs, respectively, where the 5' two bases in the tRNA anticodons complement the 5' two bases of the corresponding codons (16). Leucine tRNA most certainly has the capability of translating each doublet base in positions a and b in the figure by the procaryotic mechanism. In fact, phenylalanine tRNA, which is also found at the frameshift site in HIV (2), can decode UU and UC doublets at this site utilizing the 5' two bases in its anticodon.

An equally plausible mechanism for frameshifting would involve triplet decoding where the base at position 1 (see fig. 1) is read in an overlapping manner (1). In the case of MMTV, for example, this base would be read once in the zero frame by Asn-tRNA and once in the -1 frame by Leu-tRNA (1). However, triplet decoding where a single base is read in an overlapping fashion has been ruled out in bacteria (17); and therefore, at present, the doublet decoding model discussed above may be considered the more likely one. The possibility that a "slippage" may occur where the codon at positions a, b and c is translated and then the reading frame slips back one base to establish the -1 frame (2,3) seems unlikely for the reasons given above to rule out the base in position c as part of the frameshift site. Additionally, the possibility that doublet decoding of positions 3 and 2 may occur to achieve the -1 reading frame seems unlikely since the 5' two bases in the asparagine anticodon do not complement the AA residues in the AAC codon (see the homopolymeric signal used in MMTV, HTLV-1 and BLV in fig. 1).

The codons immediately upstream of the frameshift site (i.e., those decoded by bases 3, 2 and 1 in the figure) correspond in many cases to tRNAs which normally contain a hypermodified base in the anticodon loop. For example, Q base occurs in Asn-, Asp-, His- and Tyr-tRNAs and Wye base in Phe-tRNA. The coding properties of tRNAs lacking Q (18-20) or Wye base (21) in their anticodon loop are altered. Furthermore, it seemed reasonable that the frameshift event, whether it occurs by doublet decoding or by overlapping reading of the base designated 1 in the figure, may be facilitated if the tRNA decoding bases 3, 2 and 1 does not have a highly modified base in the anticodon loop. It seemed reasonable therefore to examine the chromatographic properties of aminoacyl-tRNAs at and around the frameshift site. For this purpose, tRNA was extracted from HIV-1, BLV and HTLV-1 infected cells and from a corresponding set of uninfected (control) cells. HIV-1 utilizes Phe-tRNA (UUU) just before the gag-pol frameshift site and BLV and HTLV-1 utilize Asn-tRNA just before the gag-pro frameshift site. tRNA was aminoacylated with radioactive amino acid and the chromatographic properties of labeled aminoacyl-tRNAs from infected and uninfected cells compared on a reverse phase chromatographic column (22). Comparison of the chromatographic properties of Asn-, Phe-, Leu- and Arg-tRNAs from HIV-1 infected and uninfected cells (note that each tRNA is utilized in the HIV frameshift sequence in fig. 1) showed that: 1) virtually all of the Asn-tRNA from infected cells was Q deficient, while a greater proportion from uninfected cells was Q containing; 2) virtually all of the Phe-tRNA from infected cells was Wye deficient, while most of the Phe-tRNA from uninfected cells was Wye containing; and 3) Leu- and Arg-tRNAs from infected and uninfected cells were very similar. The chromatographic properties of Ala-, Gly-, Lys-, Ser- and Val-tRNAs, which have been reported to be altered in other malignant cells and tissues (see review in 23), were also examined. No alterations in the chromatographic properties of these aminoacyl-tRNAs in infected and uninfected cells were observed with the exception of Lys-$tRNA_3$. Interestingly, Lys-$tRNA_3$ is the primer tRNA for HIV reverse transcriptase (9,10). We are presently examining the possible relationship of the alteration in Lys-$tRNA_3$ and its ability to serve as primer for HIV DNA synthesis.

The chromatographic properties of Ser-, Lys-, Asn-, Leu-, Phe-, Pro- and Leu-tRNAs from BLV infected and uninfected cells (note that these tRNAs are utilized in BLV gag-pro and pro-pol frameshift sequences in fig. 1) were also compared. The results showed that: 1) virtually all of the Asn-tRNA from infected cells was Q deficient, while a greater proportion from uninfected cells was Q containing; and 2) the other tRNAs from infected and uninfected cells were very similar. The chromatographic properties of each of the aminoacyl-tRNAs utilized in translating HTLV-1 gag-pro and pro-pol sequences shown in fig. 1 were also examined. A control set of uninfected cells for HTLV-1 was not available. The results showed that virtually all of the Asn-tRNA was Q deficient, while the chromatographic properties of Pro-, Ser-, Leu- and Phe-tRNAs were similar to those observed in BLV infected and uninfected cells.

We also compared the levels of aminoacylation of 12 tRNAs (Ala-, Arg-, Asn-, Asp-, Gly-, His-, Leu-, Lys-, Phe-, Ser-, Tyr- and Val-tRNAs) from HIV-1 infected and uninfected cells. The data show that the amounts of individual tRNAs from infected cells were similar to those observed in uninfected cells. This study provides evidence that the undermodified isoacceptors from HIV infected cells arise from alterations occurring in the expressed tRNA population and not from induction of additional tRNA species.

In summary, Asn- and Phe-tRNAs were hypomodified in HIV infected cells and these tRNAs are required for translation of codons immediately upstream of the frameshift site. Similarly, Asn-tRNA was hypomodified in BLV and HTLV-1 infected cells and Asn-tRNA is required for translation of the AAC codon which occurs immediately upstream of each respective frameshift site. As noted above, the coding properties of tRNAs lacking Q base (18-20) and Wye base (21) are altered. Presumably, the presence of G in place of the hypermodified Q base in the 5' position of the anticodon of Asn-tRNA (18-20), or of 1-methylG in place of the hypermodified Wye base in the 3' position next to the anticodon of Phe-tRNA (24), would facilitate a frameshift involving a two base translocation of the bases designated a and b in fig. 1 or possibly a frameshift involving an overlapping reading of the base designated 1. The absence of the large, highly modified Q base in the wobble position of the anticodon of Asn-tRNA would yield more space in and around the base at position 1. The tRNA causing the frameshift presumably could more readily "shift" its anticodon so that the 5' two bases of the anticodon decode positions a and b (i.e., doublet decoding) and the remainder of the anticodon loop (including the 3' position) is accommodated by the occurrence of more space in and around the base designated 1. Alternatively, the frameshift tRNA presumably could more easily decode base 1 by overlapping reading due to greater accessibility created by the absence of the highly modified Q base. Although the highly modified Wye base is on the 3' side of the anticodon of Phe-tRNA, it can be argued that greater movement in the anticodon of Phe-tRNA is created by absence of Wye base providing more space in and around the base at position 1 for the frameshift to occur. A search of the frameshift sequences in retroviruses (4-14) shows that the tRNA which is required for translation of the codon immediately upstream of the frameshift site is one that normally contains Q base or is Phe-tRNA. Interestingly, the only exception to this observation is the occurrence of a proline codon (CCU) in positions 3, 2 and 1 at the pro-pol frameshift signal in HTLV-1 and BLV (see fig. 1). Mammalian $tRNA^{Pro}$ which reads the codon CCU lacks a hypermodified base in its anticodon loop (25). Thus, it seems that tRNAs which decode the site designated by bases 3, 2 and 1 in fig. 1 do not contain a hypermodified base in the anticodon loop. It is also of interest to note that leucine isoacceptors, which are required for translation of codons at the frameshift site in many retroviruses, lack a hypermodified base in their anticodon loop (see 26 and references therein). In light of the observations made in this study, it is tempting to speculate that lack of a hypermodified base in the anticodon loop of tRNAs which decode bases 3, 2 and 1 and bases a and b is a requirement for the frameshift event. In addition, the occurrence of two signals for the frameshift event may suggest that different leucine isoacceptors are utilized in frameshifting; i.e., one isoacceptor responds to the UUUA signal and another to the AAAAAAC signal. Furthermore, if the changes in the tRNA populations of infected cells observed in this study are essential to viral expression, then these observations may provide an avenue for altering viral expression (27).

Acknowledgements

The authors express their appreciation to Dr. L. Arthur for the HIV-1 infected H9 cells and the control H9 cells and to Drs. J. Kern and M. Sarngadharan for the HTLV-1 infected MT-2 cells. This study was supported in part by NCI under contract #NO1-CO-74101 with Bionetics Research, Inc.

References

1. Hizi, A., L.E. Henderson, T.D. Copeland, R.C. Sowder, C.V. Hixson, S. Oroszlan. 1987. Proc.Natl.Acad.Sci.USA 84, 7041.

2. Jacks, T., M.D. Power, F.R. Masiarz, P.A. Luciw, P.J. Barr, H.E. Varmus. 1988. Nature 331, 280.

3. Jacks, T., K. Townsley, H.E. Varmus, J. Majors. 1987. Proc.Natl.Acad. Sci.USA 84, 4298.

4. Moore, R., M. Dixon, R. Smith, G. Peters, C. Dickson. 1987. J.Virol. 61, 480.

5. Rice, N.R., R. Stephens, A. Burny, R. Gilden. 1985. Virology 142, 357.

6. Sagata, N., T. Yasunaga, J. Tsuzuku-Kawamura, K. Ohishi, Y. Ogawa, Y. Ikawa. 1985. Proc.Natl.Acad.Sci.USA 82, 677.

7. Inoue, J.-I., T. Watanabe, M. Sato, A. Oda, K. Toyoshima, M. Yoshida, M. Seiki. 1986. Virology 150, 187.

8. Shimotohno, K., Y. Takahashi, N. Shimizu, T. Gojobori, D.W. Golde, I.S. Chen, M. Miwa, T. Sugimura. 1985. Proc.Natl.Acad.Sci.USA 82, 3101.

9. Ratner, L., W. Haseltine, R. Patarca, K.J. Livak, B. Starcich, S.F. Josephs, E.R. Doran, J.A. Rafalski, E.A. Whitehorn, K. Baumeister, L. Inanoff, S.R. Petteway, M.L. Pearson, J.A. Lautenberger, T.S. Papas, J. Ghrayeb, N.T. Chang, R.C. Gallo, F. Wong-Staal. 1985. Nature 313, 277.

10. Wain-Hobson, S., P. Sonigo, O. Danos, S. Cole, M. Alizon. 1985. Cell 40, 9.

11. Guyader, M., M. Emerman, P. Sonigo, F. Clavel, L. Montagnier, M. Alizon. 1987. Nature 326, 662.

12. Franchini, G., C. Gurgo, H.-G. Guo, R.C. Gallo, E. Collati, K.A. Fragnoli, L.F. Hall, F. Wong-Stall, M.S. Reitz. 1987. Nature 328, 539.

13. Chakrabarti, L., M. Guyader, M. Alizon, M.D. Daniel, R.C. Desrosiers, P. Tiollais, P. Sonigo. 1987. Nature 328, 543.

14. Schwartz, D.E., R. Tizard, W. Gilbert. 1983. Cell 32, 853.

15. Jacks, T., H.E. Varmus. 1985. Science 230, 1237.

16. Dayhuff, T., J. Atkins, R. Gesteland. 1986. J.Biol.Chem. 261, 7491.

17. Bruce, A.G., J. Atkins, R. Gesteland. 1986. Proc.Natl.Sci.USA 83, 5062.

18. Meier, F., B. Suter, H. Grosjean, G. Keith, E. Kubli. 1985. EMBO J. 4, 823.

19. Bienz, M., E. Kubli. 1981. Nature 294, 188.

20. Beier, H., M. Barciszewska, H.D. Sickinger. 1984. EMBO J. 3, 1096.

21. Smith, D.W.E., D. Hatfield. 1986. J.Mol.Biol. 189, 663.

22. Hatfield, D., C.R. Matthews, M. Rice. 1979. Biochim.Biophys.Acta 564, 414.

23. Nishimura, S. 1979. In: Transfer RNA: Structure, properties and recognition (Part A) (P. Schimmel, D. Soll, J. Abelson, eds.). p. 59.

24. Kuchino, Y., E. Borek, D. Grunberger, J. Mushinski, S. Nishimura. 1982. Nuc.Acids Res. 10, 6421.

25. Harada, F., G. Peters, J.E. Dahlberg. 1979. J.Biol.Chem. 21, 10979.

26. Valle, R.P.C., M.-D. Morch, A.-L. Haenni. 1987. EMBO J. 6, 3049.

27. Hatfield, D. 1985. TIBS 10, 201.

BIOCHEMICAL CHARACTERIZATION OF RETROVIRUS PROTEASE

Y. Yoshinaka, I. Katoh, M. Adachi
Japan Immunoresearch Laboratories Co., Ltd.
Takasaki, Gunma, 370 Japan

N. Yamamoto
Yamaguchi University School of Medicine
Ube, Yamaguchi, 755 Japan

A. Ikai
Department of Biophysics and Biochemistry
University of Tokyo, Hongo, Tokyo, 113 Japan

Y. Ikawa
Laboratory of Molecular Oncology
Tsukuba Life Science Center
Institute of Physical and Chemical Research
Tsukuba, Ibaraki, 305 Japan

Introduction

The replication of many animal and plant viruses is entirely dependent on proteolytic processing, and proteolytic enzymes are involved in a variety of functions including separation of structural and non-structural proteins, generation of specific enzymes (e.g., RNA polymerase), coordinated assembly of the virion, and particle maturation (1,2). In the retrovirus system, all replication competent viruses encode a protease in the gag-pol junction in the viral genome, which is expressed either by readthrough or frame-shift suppression of the gag termination codon. (In the avian system, protease is encoded by the gag gene; 3,4,5) As summarized in Fig. 1, precursor polyproteins of viral components such as gag and gag-pol (Pr^{gag} and $Pr^{gag\text{-}pol}$), and viral RNA are assembled just underneath the

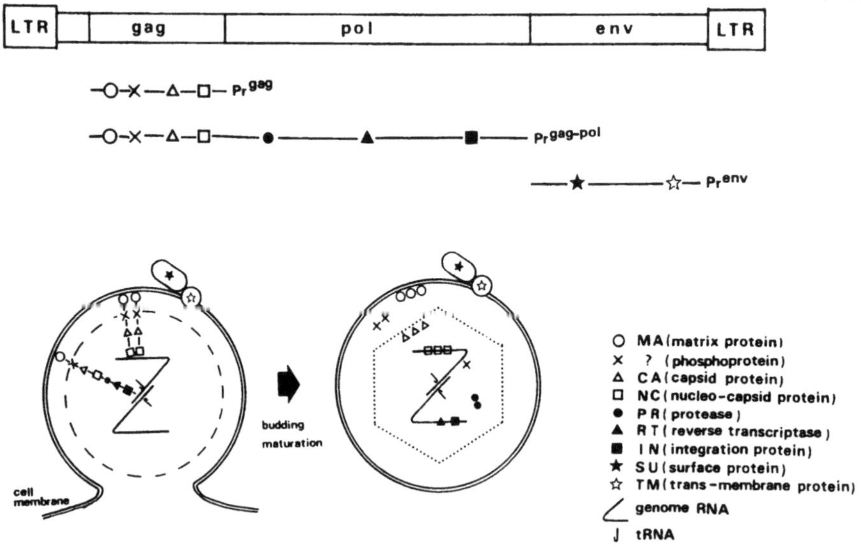

Fig. 1 A model for retrovirus particle assembly and maturation

cellular membrane where the virus buds. During this step the viral protease, incorporated as a part of the gag-pol polyprotein and most likely released autocatalytically, cleaves the precursors into mature core components and enzymes, accompanied by a morphological conversion of the core structure from "immature" to "mature" forms. The protease-deficient particles obtained by in vitro mutagenesis abolish cleavage of the precursor proteins, and result in formation of "immature" non-infectious particles (6). Thus retrovirus protease plays an important role in the production of "mature" and infectious virus particles.

To understand the proteolytic mechanism of viral protease, we studied the biochemical properties of retroviral proteases.

Results

I. Purification of the Retrovirus Proteases

Purification of the protease from purified Moloney murine leukemia virus (MuLV), bovine leukemia virus (BLV), and avian myeloblastosis virus (AMV) was done as described earlier (3,4,7). As shown in Fig. 2, we employed reverse-phase high pressure liquid chromatography (RP-HPLC) as the final step in the procedure in order to obtain pure material without loss of protease activity.

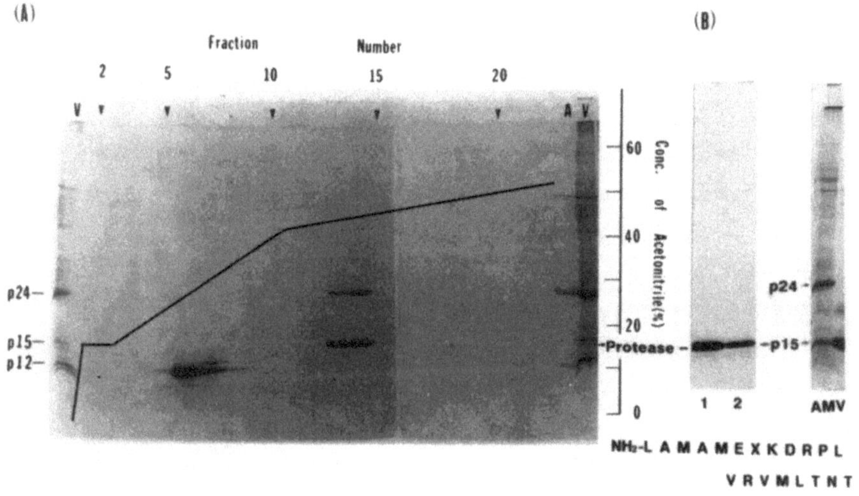

Fig. 2 Purification of BLV and AMV proteases. (A) Isolation of BLV protease by RP-HPLC with acetonitrile gradient. BLV protease was eluted in fractions 17 and 18. The acetonitrile gradient is shown. (B) Two preparations of purified AMV protease are shown in lanes 1 and 2. NH_2-terminal amino acid sequences determined with these preparations are also indicated.

II. Are Retrovirus Proteases Active in Dimeric Form?

Amino acid sequence analysis revealed that retrovirus proteases contain a highly conserved region which is similar to the sequence around the aspartic acid residues of the active site of aspartic proteinases (8,9). Although aspartic proteinases are two-domain enzymes more than 300 residues in length, with a conserved sequence in each domain, retrovirus proteases are 99 to 126 amino acid residues in length and contain only one conserved sequence, -Asp-Thr-Gly-. These observations suggest that formation of dimers could make the retroviral enzymes equivalent to general aspartic proteinases (9,12). To test this possibility several experiments were performed.

(i) Gel filtration and ultrafiltration

To determine the apparent molecular weight (MW) of BLV and AMV proteases, crude extracts containing the proteases were fractionated by Sephadex G-75 column chromatography. Both proteases were eluted in a broad peak around positions from MW 14K to 30K. The major peak was observed at 30K position showing dimers, indicating that the proteases are composed mostly of dimers with a small proportion of monomers.

(ii) Ultracentrifugal analysis

Ultracentrifugal analysis was performed with HPLC-purified AMV protease (Fig. 3). The MW calculated from the slope was 22,800 when the partial specific volume of the molecule was assumed to be 0.73 cm^3/g. Since the MW of the protease polypeptide is about 14,000, the result of sedimentation equilibrium suggested that the solution contained mostly a mixture of monomers and dimers in a weight ratio of 0.4:0.6.

(iii) Chemical cross-linking

AMV and BLV proteases were reacted with various concentrations (10^{-3} to 10^{-6} M) of a cross-linking reagent, disuccinimidyl suberate (DSS) in phosphate buffer (pH6.5). After 4 hours incubation at 24°C, the reactions were analyzed by SDS-PAGE.

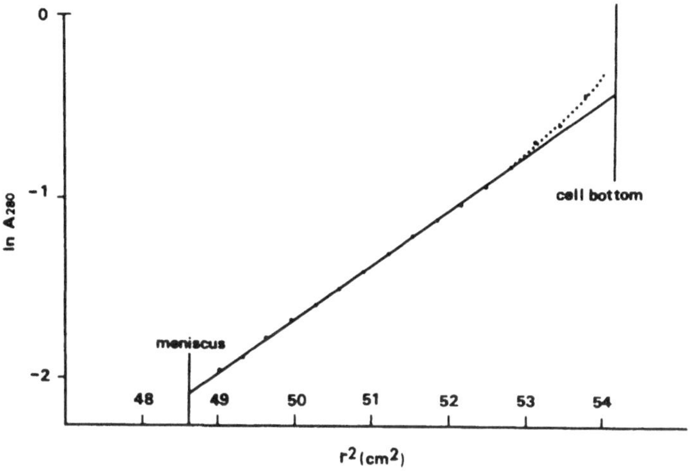

$$\sigma = \frac{\Delta y}{\Delta x} = 0.305$$

$$M(1-\bar{v}\rho) = \sigma \times \frac{2RT}{\omega^2} = 6162$$

$$\text{when} \quad \bar{v} = 0.73 \quad \rho = 1.00\,, \qquad M \sim 22800$$

Fig. 3 The result of sedimentation equilibrium experiment. The absorption at 280nm of the solution in the centrifuge cell was plotted as a function of the square of the distance from the rotor center, r^2. The slope of the plot gives $M(1-\bar{v}\rho)^2/2RT$ where M, $\bar{v}$, ρ, R, and T are molecular weight, partial specific volume, specific density, angular velocity, gas constant, and temperature in K, respectively.

The patterns for both proteases are typical for protein molecules composed of two identical subunits: a 30K protein was produced as a cross-linked product without any oligomer formation. This result too supports the theory of dimer formation of retrovirus protease molecules.

(iv) Inhibition of the protease activity by fragmented protease

If retrovirus protease is active in dimeric form, peptide fragments obtained from the protease molecules should inhibit the activity by acting as false counterparts of the entire molecules. To examine this possibility, CNBr-fragmented AMV protease, a mixture of six fragments, was added into the AMV protease assay reaction. As expected, 2µg, a 20-fold excess amount of active protease, suppressed the activity by 50%. In a control experiment, a 200-fold excess amount of CNBr-fragmented bovine serum albumin (BSA) had no effect on protease activity.

These lines of evidence strongly suggest that retrovirus proteases are active in dimeric form. If this is true, it raises an interesting question: how do homo-dimeric proteases with two identical aspartic acid residues in the active site catalyze the proteolytic reaction? This may be answered when detailed structural analysis of these proteases becomes available by means of crystallography.

III. Retroviral Proteases Belong to the Aspartic Proteinase Group

As mentioned above, the retroviral enzymes and aspartic proteinases share the conserved sequence -Asp-Thr-Gly-, which is known as the amino acid residue of the active site in aspartic proteinases. To examine whether retrovirus proteases have biochemical properties related to aspartic proteinases, the following experiments were made.

(i) pH-dependency and reaction time

Purified BLV protease and substrate, and murine sarcoma virus-derived gag-polyprotein precursor $Pr65^{gag}$ (14), were incubated under various acetate buffered pH conditions (pH 3.0 to 5.5) or with PIPES (pH 5.5 to 8.0), for 16 hours at 22°C. As shown in Fig. 4, BLV protease was most active between pH 5.5 and 6.5, as judged by the decrease of $Pr65^{gag}$ band intensity appearing on the SDS-PAGE gel. Interestingly, the cleavage profiles of $Pr65^{gag}$ were altered by pH level. This could be explained by a different protease-susceptible site being exposed due to the

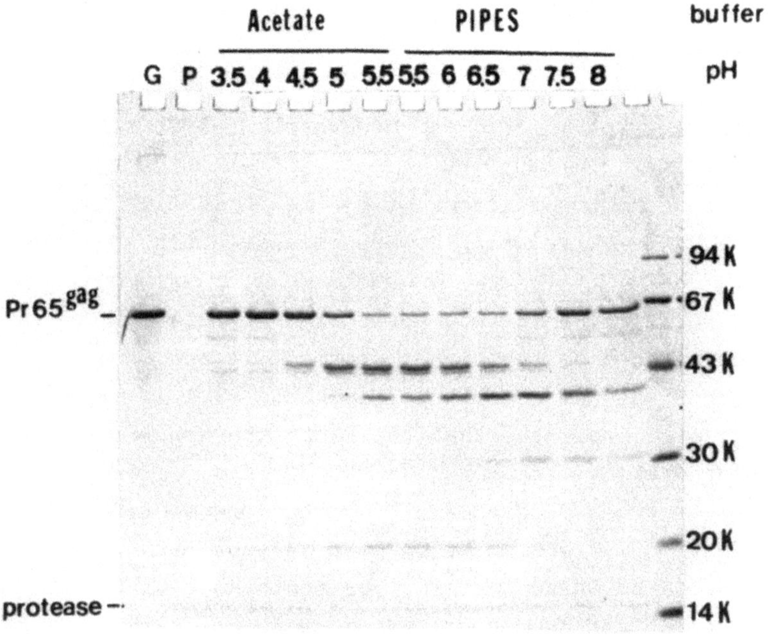

Fig. 4. Cleavage profile of Pr65gag by BLV protease under various pH conditions. HPLC-purified BLV protease (0.2µg) was incubated with Pr65gag substrate (3µg) in the buffers indicated. Reactions were stopped and Pr65gag processing was analyzed by SDS-PAGE.

conformational change caused by pH alteration. Products of 30K and 40K, obtained at higher pH (7.0 to 8.0), have the same antigenicity, respectively, as p30 and p40 generated by MuLV protease. The 43K product detected at lower pH (5.5 to 7.0) is not observed in the MuLV protease. Similar pH dependencies with respect to maximum activity were observed in other proteases of AMV, MuLV, human T-cell leukemia virus (HTLV), and human immunodeficiency virus (HIV).

Retrovirus proteases need unusually long incubation times to achieve sufficient cleavage: 5 to 16 hours at 22°C. (At 37°C, reactions for 16 hours yield 70% of the cleavage obtained at

22°C.) These observations thus reveal an unusual feature of the virus proteases.

(ii) Substrate specificity and cleavage sites in $Pr65^{gag}$

When whole virus preparations of Nonidet P-40 (1%) -disrupted MuLV, BLV, AMV, HTLV, and HIV were incubated with $Pr65^{gag}$ as the substrate, only $Pr65^{gag}$ and its related or intermediate cleavage products were processed into their final products; none of the many contaminated or associated proteins in the virus preparations was changed after incubation (3,4,7, Fig.6). This result clearly indicates that the substrate specificity for viral proteases is very high. We tested many purified proteins and peptides, including casein, bovine serum albumin (BSA), ovalbumin, fibrin, and specific peptide substrates used in various protease assay systems, for their suitability as substrates for retrovirus proteases. None of these materials was cleaved under non-denaturing conditions. The B-chain of insulin, obtained by carboxamidomethvlation and denaturation of insulin by SDS, can be cleaved by AMV protease as reported earlier (13). Assay systems using synthetic peptide substrates, however, might be useful for accurate and quantitative detection of activity.

$Pr65^{gag}$ is the most useful viral protease substrate so far examined. Figure 5 shows cleavage patterns of $Pr65^{gag}$ by retrovirus proteases of various origins. The results suggest that avian and mammalian retrovirus proteases may recognize different $Pr65^{gag}$ molecule sites. However, any contaminated protease present might possibly cleave the $Pr65^{gag}$. Nevertheless, due to its sensitivity to a variety of retrovirus proteases, $Pr65^{gag}$ could serve as a common and convenient substrate even in assays with crude protease preparations.

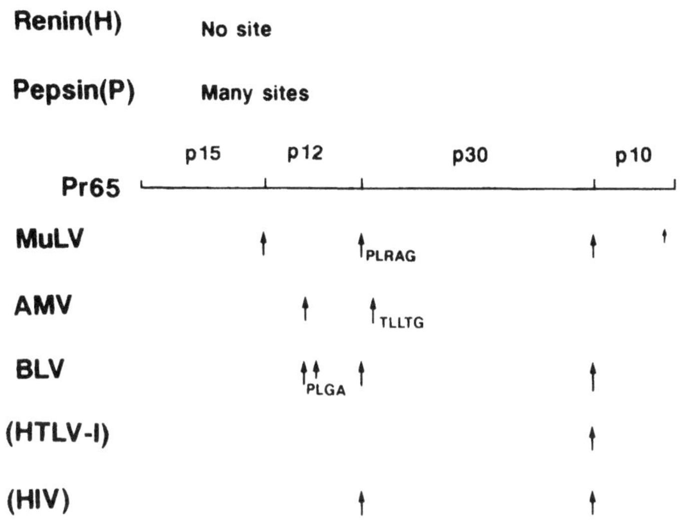

Fig. 5 Cleavage of $Pr65^{gag}$ by retrovirus proteases of various origins. Relative positions of the cleavable sites, as determined by immunoblot analysis of the products, are indicated by arrows. NH_2-terminal amino acid sequences determined after purification of the products are also indicated. Purified proteases of MuLV, AMV and BLV were incubated with the substrate in 20mM PIPES buffer (pH6.5) for 16 hrs at 22°C. In the experiments with HTLV-1 and HIV, virus particles were disrupted with NP-40 and incubated with $Pr65^{gag}$ under the same conditions.

(iii) Effect of various protease inhibitors against retrovirus proteases.

Several protease inhibitors were introduced into a protease assay system with purified BLV protease. Among many inhibitors tested, pepstatin A, a specific inhibitor of aspartic proteinases, was most effective (Table 1) (9). Yet the 50% inhibitory concentration of pepstatin A against BLV protease was very high ($IC_{50}=2\times10^{-4}$ M) in comparison to general aspartic proteinases such as porcine pepsin ($IC_{50}=10^{-8}$ M).

Another compound, diazoacetyl norleucine methyl ester (DAN), which esterifies the active site of aspartic residues of aspartic proteinases, also inactivated BLV and AMV proteases in the presence of Cu^{++}. From these results, and from the amino acid sequence homology between retrovirus and aspartic proteinases, we conclude that retrovirus proteases belong to the aspartic proteinase group. Nevertheless, the substrate specificity and molecular organization of the retrovirus proteases, as well as their sensitivity to pepstatin A, differentiate them from general aspartic proteinases. Hence retrovirus proteases can be identified as a new group of aspartic proteinases.

Table 1 Effects of Various Protease Inhibitors on Purified BLV Protease.

Inhibitor	Specificity	IC_{50} (mM)
pepstatin A	aspartic proteinase	0.2
soybean trypsin inhibitor	serine proteinase (trypsin)	>1
TLCK	serine proteinase (trypsin)	>1
leupeptin	serine proteinase (trypsin)	>3
aprotinine	serine proteinase (trypsin)	>2
chymostatin	serine proteinase (chymotrypsin)	>1
TPCK	serine proteinase (chymotrypsin)	>1
antipain	cysteine proteinase	>1
PCMS	cysteine proteinase	>2
N-ethylmaleimide	cysteine proteinase	>8
E-64	cysteine proteinase	>3
EDTA	metallo-proteinase	>1
phosphoramidon	metallo-proteinase	>2

TLCK, tosylsulfonyl-lysyl chloromethyl ketone; TPCK, tosylsulfonyl-phenylalanyl chloromethyl ketone; PCMS, p-chloromercuriphenylsulfonic acid; EDTA, ethylenediamine-tetraacetic acid

Finally, as shown in Figure 6, Western blot analysis revealed that HIV protease is also inhibited by pepstatin A: increasing concentrations of pepstatin A inhibited the cleavage of Pr65^{gag} as detected by the increasing band intensity of Pr65^{gag} in the stained gel. The cleavage products of 40K and 30K indicated by arrows have antigenic determinants of p30+p10 and p30 of MuLV gag protein, respectively. Thus HIV protease appears to cleave Pr65^{gag} with a specificity similar to that of MuLV protease, and its activity is inhibited by pepstatin A.

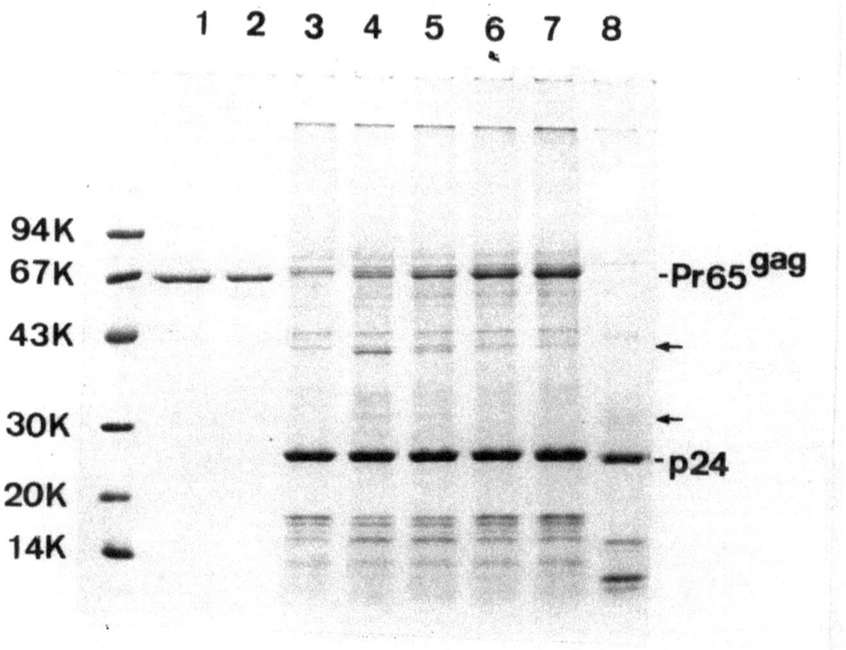

Fig. 6 SDS-PAGE gel showing the effect of pepstatin A on HIV protease activity. The substrate Pr65^{gag} and HIV proteins are shown in lanes 1 and 3, respectively. Detergent-disrupted HIV (10µg) was incubated with 2µg of Pr65^{gag} in the presence of pepstatin A at 0mM (lane 4), 0.1mM (lane 5), 0.5mM (lane 6) and 1mM (lane 7). Lane 2 shows Pr65^{gag} protein incubated in buffer alone. Lane 8 shows BLV proteins.

Further characterization of retrovirus proteases will allow us to find more specific and effective inhibitors. Many human retroviruses are causative agents of significant diseases; it is to be hoped that pharmacological agents which block virus particle maturation will eventually be developed to treat these diseases.

References

1. Kräusslich, H.-G., E. Wimmer. 1988. Annu. Rev. Biochem. 51, p. 107.

2. Weiss, R., N. Teich, H. Varmus, J. Coffin. 1982. Molecular Biology of Tumor Viruses: RNA tumor viruses, 2nd ed., Cold Spring Harbor, New York.

3. Yoshinaka, Y., I. Katoh, T.D. Copeland, S. Oroszlan. 1985. Proc. Natl. Acad. Sci. U.S.A. 82, 1618.

4. Yoshinaka, Y., I. Katoh, T.D. Copeland, S. Oroszlan. 1986. J. Virol. 57, 826.

5. Jacks, T., K. Townsley, H.E. Varmus. 1987. Proc. Natl. Acad. Sci. U.S.A. 84, 4298.

6. Katoh, I., Y. Yoshinaka, A. Rein, M. Shibuya, T. Odaka, S. Oroszlan. 1985. Virology 145, 280.

7. Yoshinaka, Y., R.B. Luftig. 1977. Cell 12, 709.

8. Toh, H., M. Ono, K. Saigo, T. Miyata. 1985. Nature 315, 691.

9. Katoh, I., T. Yasunaga, Y. Ikawa, Y. Yoshinaka. 1987. Nature 329, 654.

10. Power, M.D., P.A. Marx, M.L. Bryant, M.B. Gardner, P.J. Barr, Lucino. 1986. Science 231, 1567.

11. Foltmann, B., V.B. Pederson. 1977. In: Acid Proteases: Structure, Function and Biology (Tang, J., ed.) Plenum Press, New York, p. 3.

12. Pearl, L.H., W.R. Taylor. 1987. Nature 329, 351.

13. Dittmar, K.J., K. Moelling. 1978. J. Virol. 28, 106.

14. Yoshinaka, Y., R.B. Luftig. 1982. Virology 118, 380.

15. Kotler, M., R.A. Katz, W. Danho, J. Leis, A.M. Skalka. 1988. Proc. Natl. Acad. Sci. U.S.A. 85, 4185.

EXPRESSION AND PROCESSING OF p15^{gag} PROTEASE OF MYELOBLASTOSIS ASSOCIATED VIRUS IN *E. COLI*

J. Sedláček, F. Kaprálek

Department of Gene Manipulations, Institute of Molecular Genetics, Czechoslovak Academy of Sciences; 166 37 Prague 6

P. Štrop, I. Pichová, V. Kostka

Department of Biochemistry, Institute of Organic Chemistry and Biochemistry, Czechoslovak Academy of Sciences; 166 10 Prague 6

M. Trávníček

Department of Molecular Virology, Institute of Molecular Genetics, Czechoslovak Academy of Sciences; 166 37 Prague 6

Introduction

Retroviral proteases are typically encoded by the 5' region of the *pol* open reading frame in mammalian retroviruses or by the 3' region of the *gag* open reading frame in avian retroviruses [1]. The myeloblastosis associated virus (MAV) is a replication-competent helper retrovirus [2] of an acute defective avian retrovirus, the avian myeloblastosis virus (AMV). The coding stretch for the protease of MAV, p15^{gag}, comprises 124 codons in the 3' region of the *gag* open reading frame and is followed by a stop codon; the protein maps at the carboxy terminus of the *gag* polyprotein precursor [3, 4]. Altogether five major structural MAV proteins are parts of the *gag* polyprotein precursor: p19, p10, p27, p12, and p15 [5, 6]. The p15^{gag} protease is supposed to cleave these proteins specifically from the common *gag* precursor as well as to cleave the p63/p32 proteins from the *gag-pol* precursor.

Proteases of Retroviruses

The release of two major *env* proteins is supposed to be catalyzed by a host protease of serine type. The protease of the MAV is identical to that of the Rous sarcoma virus except for amino acids in three positions: 52 (T in RSV, A in MAV), 54 (W in RSV, R in MAV), and 82 (L in RSV, V in MAV) [3, 4]. The recovery of these proteases from virions in infected cells is often limited and their isolation laborious [7]. The recent recourse to recombinant-DNA expression in heterologous systems viz. *E. coli* [8, 9] and yeast [10], has led to remarkable progress in our understanding of several aspects of the formation of retroviral proteases as well as of some of their protein and enzyme properties. Below we describe the obtainment of MAV p15*gag* protease from a very efficient *E. coli* system expressing a 24 kD 3' end portion of the *gag* open reading frame that encodes 65 amino acids of the p12*gag* protein and the complete p15*gag* protease [11]. By separation of the expression product from inclusion bodies, the correctly processed, active p15*gag* protease was obtained in a high yield. Other cleavage products of the primary translate were also identified. The formation of an extended p15*gag* molecule containing 28 amino acids of p12*gag* at the N-terminus appears to be the initial event of the activation of p15*gag* from its precursor.

Results and Discussion

To obtain a massive supply of the MAV p15*gag* protease, a 0.82 kb segment of MAV proviral DNA [2] comprising the 3' part of the *gag* region was manipulated into an expression plasmid construct [11]. This 0.82 kb proviral DNA segment encodes 65 amino acids of the p12*gag* moiety of the *gag* polyprotein precursor and all 124 amino acids of p15*gag*;

a translational stop signal follows (see [6] and inset of Fig. 4. for details). To the 5' end of the cloned proviral DNA fragment a segment of *E. coli* β-galactosidase DNA and a linker sequence were attached, thus extending the amino end of the viral protein by another 8 amino acids. This truncated and fused precursor of p15^{gag} was designated Pr25$^{lac\text{-}\Delta gag}$, according to its composition and calculated molecular weight. The choice of the length and structure of the recombinant precursor was a compromise based on several considerations. "Direct" expression of p15^{gag} would have changed its amino terminus owing to the presence of the initiation methionine whereas the expression of the entire natural polyprotein precursor Pr76^{gag} would have introduced additional protein components. Furthermore, doubts about the capability of Pr76^{gag} to process itself autocatalytically [5, 6, 8] did not favor the latter option. In the expression-plasmid construct employed, pGM45 [11], control elements which had proved efficient in the expression of another protease zymogen, fusion calf prochymosin [12], were used: P_{tac} and the translation initiation signal of *E. coli* β-galactosidase. Cultivations of *E. coli* strain MT harboring pMG45 were also carried out in analogy with the previous system. In induced cultures, all the cells formed polar cytoplasmic inclusions that filled a considerable part of the cell volume. The results of a typical cultivation run in a laboratory fermenter with time-delayed lactose induction are shown in Fig. 1. The biomass concentration leveled off after the eighth hour at the high value of 14.7 g of wet mass per liter; the formation of inclusion bodies began at the sixth hour and their relative volume concentration in biomass reached 12.6% by the end of the run.

The primary separation of p15^{gag} from urea-solubilized proteins of the inclusions took place during the removal of

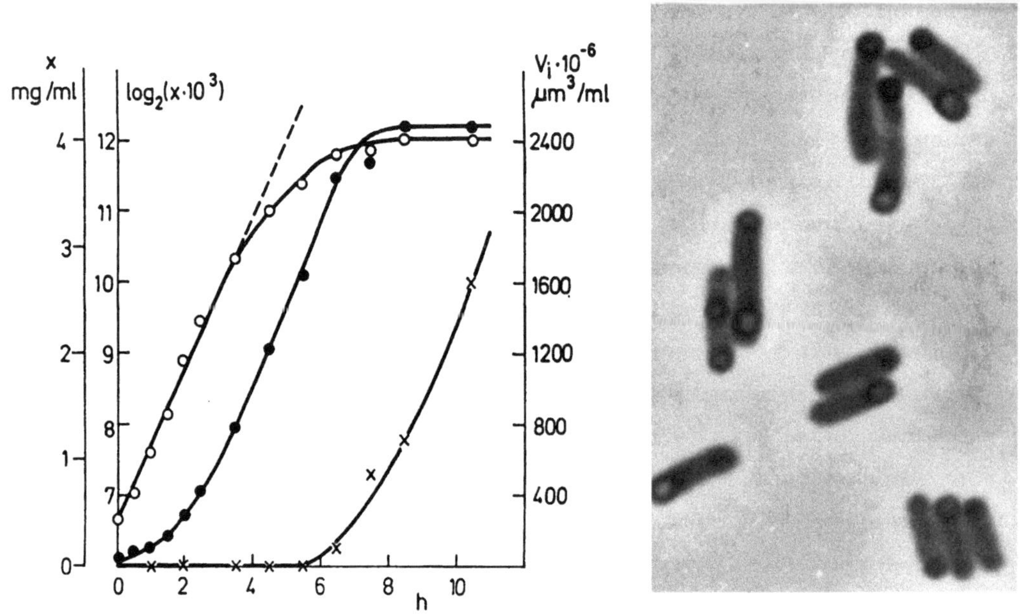

Fig. 1. Time course of cultivation of pMG45-harboring *E. coli* **MT.** - ● - Dry biomass, x, mg/ml; - ○ - $\log_2(x \cdot 10^{-3})$ - × - volume concentration of inclusions in culture, $V_i \cdot 10^{-6}$, μm^3/ml. New Brunswick Magnaferm MA 14 fermenter, filled with 8 l of medium, stirring rate 500 rpm, aeration 8 l / min. Inset: phase-contrast micrograph [13] of inclusions in cells at end of run.

urea by dialysis (Fig. 2.): the bulk of ballast proteins formed a precipitate, while the fluid above the precipitate contained nearly pure p15^{gag} in high yields. Similar partition phenomena have only rarely been mentioned for insoluble recombinant products [14]. In *E. coli* MT cells harboring pMG45, the Pr25$^{lac-\Delta gag}$ precursor formed in large quantities and was then rapidly processed to p15^{gag} protease. As was shown by Western blotting (Fig 3 A) and Coomasie blue-stained SDS-PAGE (Fig. 3 B), both the

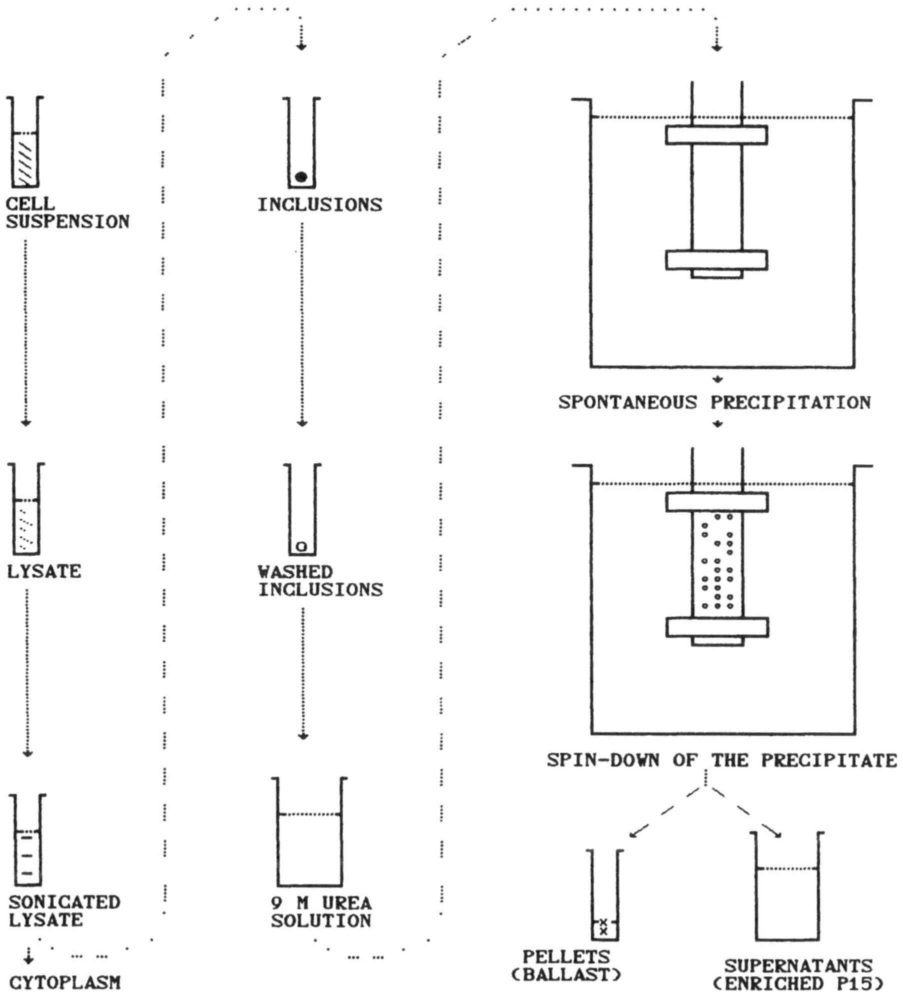

Fig. 2. Scheme of cell fractionation and separation of $p15^{gag}$ during the dialysis step. Biomass of *E. coli* MT cells harboring pMG45 was obtained as in Fig. 1. and fractionated by procedure detailed in [11, 15]. The composition of washed inclusions and of supernantant from the dialysis step is shown in Fig. 3., Lanes 2 and 4, respectively.

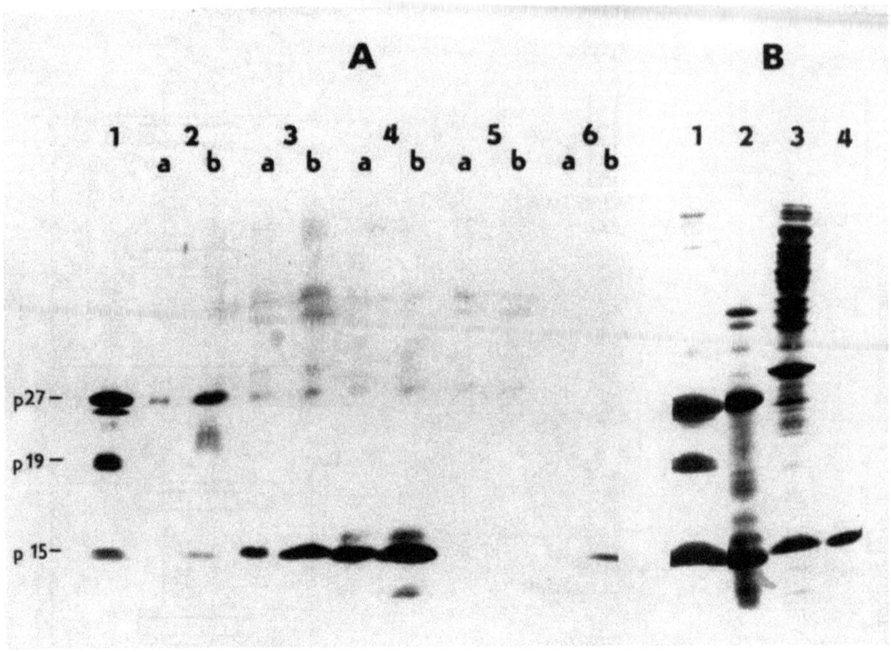

Fig. 3. Analysis of E. coli cell fractions and preparations of p15^{gag}. **A:** Immunodetection by Western blotting. Proteins bearing *gag* determinants were detected on a nitrocellulose blot of SDS-PAGE (15 % gel) by the double antibody technique using diluted rabbit antiserum directed against MAV virion *gag* proteins as the primary antibody and porcine antirabbit IgG coupled to horse-radish peroxidase as secondary antibody. Immunoprecipitation was detected with 4-chloro-1-naphthol. Lane 1: MAV virion proteins, 0.2 μg. Lane 2: inclusion bodies; 0.04 μg (a), 0.2 μg (b). Lane 3: *E. coli* cytoplasm; 0.23 μg (a), 1.15 μg (b). Lane 4: recombinant p15^{gag}, crude preparation; 0.13 μg (a), 0.65 μg (b). Lane 5: total protein of *E. coli* MT cells not harboring plasmid; 0.24 μg (a), 1.2 μg (b). Lane 6: authentic p15^{gag} from MAV virions [12]; 0.3 μg (a), 1.5 μg (b).

B: Paralel run, Coomasie blue-stained gel slab. Lane 1: MAV virion proteins, 3.4 μg. Lane 2: inclusion bodies; 16.8 μg. Lane 3: *E. coli* cytoplasm; 19.3 μg. Lane 4: recombinant p15^{gag}, crude preparation; 18.0 μg.

precursor and the protease had been incorporated into the inclusions (Fig. 3 AB; lanes 2); the cytoplasmic fraction contained processed protease only (Fig. 3 AB; lanes 3). The N-terminal amino acid sequence determined in the p15^{gag} protease purified from solubilized inclusion bodies by RP HPLC [11, 15] showed that the protease in the inclusion bodies was processed correctly. The uniform sequence L A M T M E H K found corresponded to the sequence determined in authentic MAV and RSV proteases [3, 4]. The recombinant protease from solubilized inclusion bodies was able to cleave the *gag* precursor into structural viral proteins; it also cleaved bovine and human serum albumin substrates, as shown elsewhere [11, 15]. The sequence A M T M E H was found at the N-terminus of the p15^{gag} protein isolated from the cytoplasmic fraction, after pelleting off inclusion bodies, by ion-exchange chromatography on SE-Sephadex at *p*H 5.0 and subsequent RP-HPLC on a C 3 Ultrapore column [15]; this evidenced removal of the N-terminal leucine and the presence of a high aminopeptidase activity in the cytoplasm of our *E. coli* strain, capable of cleaving the p15^{gag} molecules that had not been caught up in the inclusions.

The mechanisms and especially the first events of activation of proteases from precursors are not clear as regards their maturation both in natural host cells and in manipulated expression systems. Apparently correct processing of products of MAV proviral DNA [8] and products of fragments of ASLV and HIV proviral DNAs expressed in *E. coli* [9, 16], as well as of fragments of HIV proviral DNA expressed in yeast [10], has been reported by several authors. Structure information on the N-terminal amino acid sequence of p15^{gag} as evidence of correct processing in *E. coli* has not, however, been reported until very recently [11].

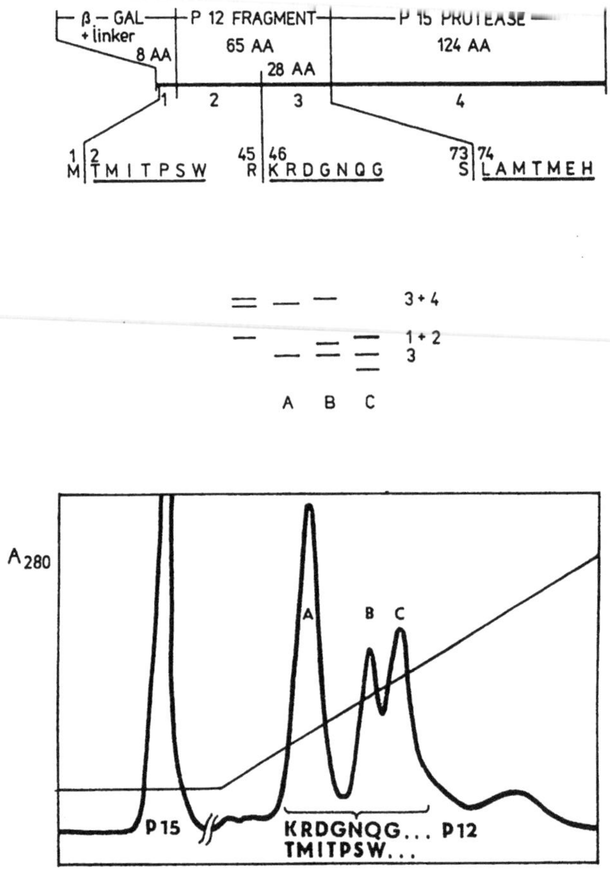

Fig. 4. N-terminal amino acid sequences found in expression products. Top: summary of alignment of sequences in $Pr25^{lac\text{-}\Delta gag}$. Center: electrophoretic mobilities of components of the chromatographic fractions A, B, C; left lane shows, from top, mobilities of reference proteins (RNAase, lysozyme, bovine pancreatic trypsin inhibitor). Bottom: elution profile; material obtained in supernatant after dialysis, at *p*H 7.1, of solubilized inclusions [11, 15] was applied to a column of SE-Sephadex equilibrated with 0.05 M phosphate *p*H 7.0, and eluted with the same buffer and with a linear gradient of 0 – 0.5 M NaCl. The N-terminal amino acid sequences were determined in the component separated, using an Applied Biosystems 470A Protein Sequencer.

We were able to separate several products of $Pr25^{lac-\Delta gag}$ processing in *E. coli* from the soluble material contained in the supernatant fraction after the preparative steps of solubilization and dialysis of inclusion bodies. Besides the product of cleavage at the main processing site, the p12-p15 boundary, other peptide products with N-terminal sequences T M I T P S W and K R D G N Q G were isolated by chromatography on SE-Sephadex at *p*H 7 (Fig. 4.) or by RP HPLC [11]. The first sequence, T M I T P S W, corresponds to *E. coli* β-galactosidase but lacks its N-terminal methionine; removal of this amino acid has been reported for several other recombinant products [17]. The other sequence found (H N A K Q L R) - K R D G N Q G, showed that processing had also taken place within the p12 structural protein region, between residues R 45 and K 46 of $Pr25^{lac-\Delta gag}$. Processing at this site appeared to be predominant, since only fragmented p12 moieties, i. e. peptides 2 and 3 shown in Fig. 4. and not the longer peptide corresponding to 2+3, were observed. Yet another species, containing 28 amino acids of p12 and the entire p15 was also found. This prolonged protein was quickly processed, and the p15 and the 28-residue fragment of p12 were formed in chromatography or under nondenaturing conditions. In contrast, we were not able to detect any proteolytic activity in precursor $Pr25^{lac-\Delta gag}$ electroeluted from SDS gels.

A sequence similar to the processing site found in the p12 part of $Pr25^{lac-\Delta gag}$, an R K R cluster of basic amino acids, has been found earlier in RSV (ALV) *env* proteins (at the p40/p23 boundary, which maps between the 405th and 406th amino acids of the *env* polyprotein) [4]. The processing of these *env* proteins is commonly believed to be catalyzed by a host protease in infected cells. It therefore seems possible that the same sequence found in the p12 part of Pr25$^{lac-\Delta gag}$ may be the site of the primary cleavage by a trypsin-type protease in natural *gag* or *gag-pol* precursors in host cells. The mechanism of release of the first catalytically active substances from precursors would then be fully understandable. Neither of the hypotheses [5, 6] assuming that whole *gag* (or even *gag-pol* precursors of most retroviruses) are catalytically active or, alternatively, that a host-cell protease provides the one and only processing for the liberation of the initial amounts of the viral protease, has been backed up by experimental evidence. Attempts to show catalytic activity of whole *gag* or *gag-pol* polyproteins *in vitro* have not been successful. A buildup of a catalytically active protease conformation at the end of the *gag* polyprotein or even in the middle of the *gag-pol* polyprotein precursors seems plausible, but if retroviral proteases acted as dimers [18] then the initial activation would require an oriented association of two large precursor molecules plus probably a third molecule of the substrate. In contrast, activation schemes postulating a conceivable step during which some quantities of short catalytically active precursors (as found for Pr25$^{lac-\Delta gag}$ in *E. coli*) are produced at the beginning of the processing by the action of a host-cell protease, appear to be simple and consistent. In general accord with this view, the mechanisms of release of active enzymes from zymogens by multiple inter- and intramolecular steps are widespread among proteases, including those of the aspartic type.

References

1. Kräusslich H.-G., E. Wimmer. 1988. *Annu. Rev. Biochem.* 57, 701-754.

2. Perbal B., J. S Lipsick, J. Svoboda, R. F. Silva, M. A. Baluda. 1985. *J. Virol.* **56**, 240-244.

3. Sauer R. T., D. W. Allen, H. D. Niall. 1981. *Biochemistry* **20**, 3784-3791.

4. Schwartz D. E., R. Tizard, W. Gilbert. 1983 *Cell* **32**, 853-869.

5. Dickson C., R. Eisenman, H. Fan, E. Hunter, N. Teich. 1984. *RNA Tumor Viruses 1* (Weiss R., N. Teich, H. Varmus, J. Coffin, eds.), pp 528-547, Cold Spring Harbor Laboratory, Cold Spring Harbor.

6. Dickson C., R. Eisenman, H. Fan. 1985. *RNA Tumor Viruses 2* (Weiss R., N. Teich, H. Varmus, J. Coffin, eds.), pp 136-146, Cold Spring Harbor Laboratory, Cold Spring Harbor.

7. Trávníček M., J. Říman. 1981. *Biochem. Biophys. Res. Commun.* **96**, 1768-1777.

8. Mermer B., M. Malamy J. M. Coffin. 1983. *Mol. Cell. Biol.* **3**, 1746-1758.

9. Kotler M., R. A. Katz, A. M. Skalka. 1988. *J. Virol.* **62**, 2696-2700.

10. Kramer R. A., M. D. Schaber, A. M. Skalka, K. Ganguly, F. Wong-Staal, E. P. Reddy. 1986. *Science* **231**, 1580-1584.

11. Sedláček J., P. Štrop, F. Kaprálek, V. Pečenka, V. Kostka, M. Trávníček, J. Říman. 1988. *FEBS Lett.* **237**, 187-190.

12. Sedláček J., M. Fábry, P. Kašpar, S. Zadražil, F. Kaprálek. 1986. *Biotechnology & Bioindustry (Sofia)* **6**, 17-19.

13. Schoemaker J. M., A. H. Brasnett, F. A. O. Marston. 1985. *EMBO Journal* **4**, 775-780.

14. Gribskov M., R. R. Burgess. 1983. *Gene* **26**, 109-118

15. Štrop P., Pichová I., Kostka V., Kaprálek F., Sedláček J. 1989. *This volume*, pp. 61-72.

16. Debouck C., J. G. Gorniak, Strickler J. E., T. D. Meek, B. W. Metcalf, M. Rosenberg. 1987. *Proc. Natl. Acad. Sci. USA* **84**, 8903-8906.

17. Marston F. A. O. 1986. *Biochem. J.* **240**, 1-12.

18. Pearl L. H., W. R. Taylor. 1987. *Nature* **329**, 351-354.

ISOLATION AND CHARACTERIZATION OF p15$^{\underline{gag}}$ PROTEASE OF MYELOBLASTOSIS ASSOCIATED VIRUS EXPRESSED IN E. coli

P. Štrop, I. Pichová, V. Kostka

Institute of Organic Chemistry and Biochemistry, Czechoslovak Academy of Science, 166 10 Prague

F. Kaprálek, J. Sedláček

Institute of Molecular Genetics, Czechoslovak Academy of Science, 166 37 Prague

Introduction

The p15$^{\underline{gag}}$ protease is the major structural protein encoded by a stretch of 124 codons in the 3' region of the gene of the myeloblastosis associated virus (MAV). Until recently this protease could be isolated only from the blood plasma of chickens with myeloblastic leukemia in its terminal stage (1); to obtain at least 10-milligram quantities of the protease blood plasma of a large number of chickens had to be collected. These quantities, however, are insufficient for any detailed investigation of the molecular and structural characteristics of the protease as well as of its specificity. A recombinant procedure was therefore developed for the production of larger quantities of the protease (2).

In this paper procedures are described for the isolation of the recombinant p15$^{\underline{gag}}$ protease expressed in E. coli and for its purification to a degree permitting its crystallization. The final preparations were characterized by RP-HPLC, FPLC and ion exchange chromatography, by SDS-PAGE and by isoelectric focusing, by amino acid analysis and N-terminal sequence analysis. The data obtained were compared with the characteristics of a sample of the authentic viral p15$^{\underline{gag}}$ protease examined simultaneously.

Results

Isolation of p15$^{\underline{gag}}$ protease from E. coli cells

Isolation from inclusion bodies

The process is shown schematically in Fig. 1. The disruption of the cells was effected by a procedure successfully employed for the production of recombinant calf prochymosin in earlier studies (3). During subsequent treatment best results were obtained when 9 M urea solutions in Tris-buffer at pH 7.1 and containing 0.1% 2-mercaptoethanol (2-ME) were used for solubilization of the inclusions and when the solution thus obtained was dialyzed against a Tris-buffer at neutral pH. When dialysis was carried out against Tris-buffer at pH 5 or when protein renaturation was attempted (by incubation of the solubilized inclusions in a large volume of buffer at pH 10.5 and subsequent neutralization) before dialysis the yields were decreased.

The high degree of purification of the p15$^{\underline{gag}}$ protease achieved when urea was removed by dialysis, a phenomenon rarely observed during the isolation of recombinant products, is remarkable. It should be noted that the inclusions contain, in addition to the protease which is the main component, also a large variety of proteins, the unprocessed precursor Pr25$^{\underline{lac-\Delta\ gag}}$ and also E.coli proteins (Fig. 2, SDS-PAGE pattern). All protein contaminants are found in the sediment formed during dialysis. The supernatant contains the correctly processed protease and small quantities of peptide and protein impurities only. A minor

amount of the protease is retained also in the sediment, most probably because of its limited solubility.
Twice repeated redissolution of the sediment followed by dialysis (Fig. 1) significantly improves the yields of the protease.

Isolation from cytoplasmic fraction

After the removal of the inclusion bodies by centrifugation (Fig. 1) the isolation of the protease from the cytoplasmic fraction was effected by one-step ion exchange chromatography on SE-Sephadex at pH 5.0 (Fig. 3).

Purification of p15$^{\underline{gag}}$protease of MAV from inclusion bodies

Ion-exchange chromatography

To remove peptide contaminants the supernatant from the dialysis step was passed over three columns used in series: a Phenyl-Sepharose column at pH 7, a SE-Sephadex column at pH 7 and a QAE-Sephadex column at pH 8 as shown in Fig. 4. The p15$^{\underline{gag}}$ protease was not retained in any of these three columns while the majority of small peptides present as contaminants were caught up in either of the ion exchange columns. The final purification of the protease was performed by chromatography on SE-Sephadex at pH 5 using the column described in Fig. 3. The total yield of the protease prepared by this procedure was about 300 mg/35 l of fermentation medium. The N-terminal amino acid sequence, amino acid composition as well as the elution profiles obtained by FPLC and RP-HPLC of the protease and its isoelectric point were identical. A sample of the protease has also been successfully crystallized (Cooper et al., and Foundling

FERMENTATION
BIOMASS
↓
Centrifugation
Lysis by Lysozyme
and Deoxycholate
Sonication
Centrifugation
↓
CYTOPLASMIC FRACTION
INCLUSION BODIES
↓
Triton wash
Centrifugation
↓
INCLUSION BODIES
↓
Solubilization
in 9M urea
3x in 0.05 M Tris
buffer, pH 7.1
Dialysis, against water
at pH 7.1 contg. 0.1% 2-ME
Centrifugation
↙ ↘
SEDIMENT **SUPERNATANT (INCLUSIONS)**

Fig. 1 Isolation of Recombinant p15$^{\underline{gag}}$ Protease of MAV

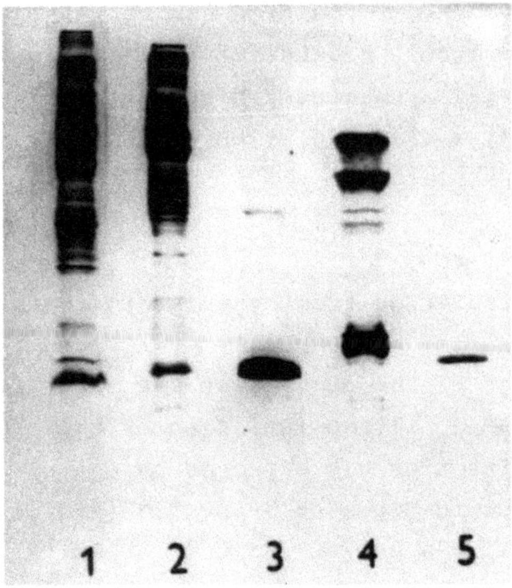

Fig. 2 SDS Page Patterns of: 1 - solubilized inclusion bodies, 2 - cell cytoplasmic fraction, 3 and 4 - supernatant and sediment resp. of solubilized inclusion bodies after 1st dialysis, 5 - recombinant p15$^{\underline{gag}}$ protease

<u>et al.</u>, unpublished observations) by using the procedure which Miller and Wlodawer had successfully employed for the homologous RSV p15$^{\underline{gag}}$ protease (kindly communicated to us by A. Wlodawer) (4).

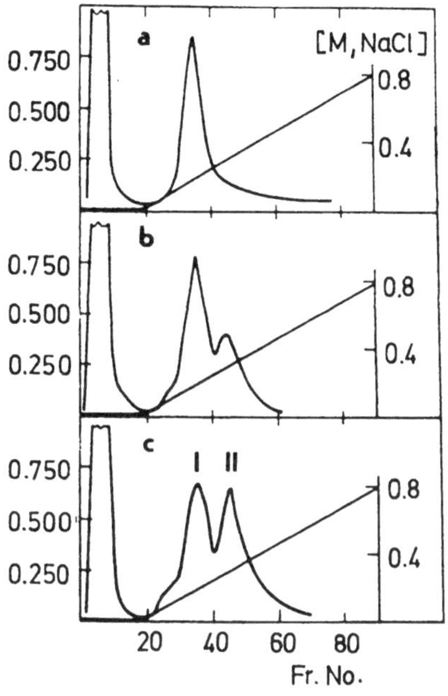

Fig. 3 Ion exchange chromatography of cell cytoplasmic fraction after a) 7 h, b) 11 h, c) 13 h of fermentation on SE-Sephadex. Peak I - protease with N-terminal sequence Ala-Met-Thr-Met-Glu-His..., peak II - peptides. Elution by 0.05 M acetate buffer, pH 5, containing 0.05% mercaptoethanol, 0.05% Triton x-100, gradient of 0-0.8 M NaCl.

Hydrophobic chromatography

A single step procedure and a Phenyl-Sepharose column equilibrated in 0.1 M Tris buffer at pH 8.5 containing 1.5 M NaCl were used. To pooled supernatants obtained by dialysis of the solubilized inclusions NaCl was added (final concentration 1.5 M) and the solution was applied to the

SUPERNATANT

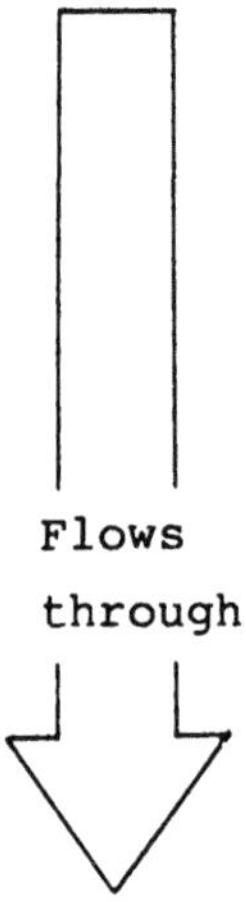

3000 ml, pH 7.1

PHENYL - SEPHAROSE CL 4B

0.02M Tris - Phosphate, pH 7.1
0.001M EDTA, 0.05% mercapto-
ethanol

SE - SEPHADEX

0.02M Tris-Phosphate, pH 7.1
0.001M EDTA, 0.05% mercapto-
ethanol
pH adjusted to 8

QAE - SEPHADEX

0.02M Tris -Phosphate, pH 8
pH adjusted to 5
Dialysis, pH 5

SE - SEPHADEX

Chromatography on SE-Sephadex

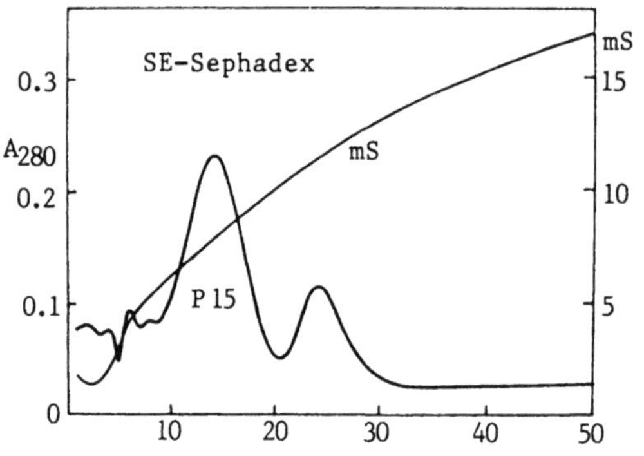

0.05M acetate, **pH 5**
0.05% mercaptoethanol
0.005% Triton X 100
Gradient 0 - 0.8M NaCl

p15 PROTEASE

Fig. 4 Isolation of p15<u>gag</u> Protease from Supernatant (after Dialysis of Solubilized Inclusion Bodies) by Ion-Exchange Chromatography column. The protease was not displaced by decreasing the ionic strength of the pH 8.5 buffer and emerged only after the application of a gradient (0-70%) of ethylene glycol (Fig. 5). The proteolytic activity of the protease was fully restored after the eluent had been removed by dialysis against 0.05 M Tris-buffer at pH 8.5. The somewhat anomalous behavior of the protease in the Phenyl-Sepharose column can be accounted for by the shielding effect of the peptide contaminants present. These peptides probably bind to the protease at pH 8.5 at low salt concentrations. At higher salt concentrations, however, hydrophobic interactions between the protease and the chromatographic support become prevalent; the peptides are then eluted while the protease remains firmly bound. Another possible explanation is that the correctly folded protease reacts only weakly with the support and that a high salt concentration has to be employed to enhance this interaction. Such concentration, however, brings about also partial unfolding paralleled by interactions which cannot be eliminated by decreasing the ionic strength.

RP-HPLC

A reverse-phase C-3 Ultrapore column in 0.05% trifluoracetic acid (TFA) in water and a gradient of 0-50% acetonitrile containing 0.05% TFA were used. The chromatography profile of a sample of the recombinant $p15^{\underline{gag}}$ protease and of a sample of the authentic viral protease is shown in Fig. 6. The proteolytic activity of the protease-containing fractions was fully restored after their neutralization.

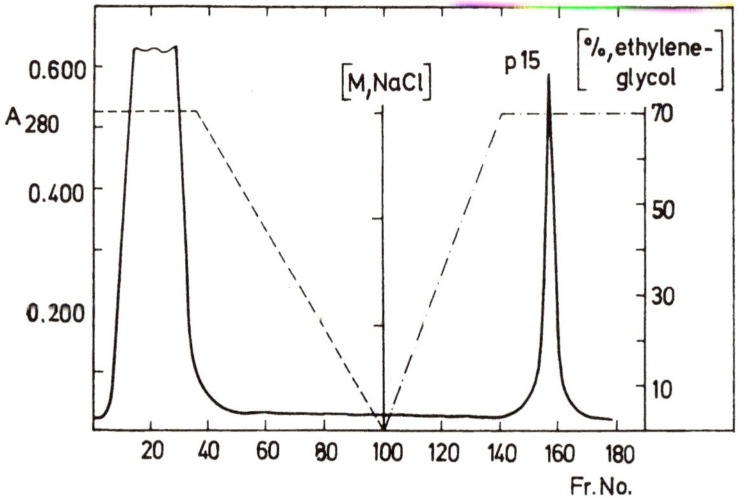

Fig. 5 Hydrophobic Chromatography of Supernatant on Phenyl Sepharose CL 4B. 50 ml of supernatant containing 1.5 M NaCl was applied to the column (70 ml) equilibrated with 0.1 M Tris buffer, pH 8.5, containing 1.5 M NaCl. Elution gradients of 1.5 - 0 M NaCl and 0 - 70% ethylene glycol.

Characteristics of Recombinant p15$^{\underline{gag}}$ Protease

Chromatographic and electrophoretic behavior

When assayed by FPLC on a Mono S column the recombinant protease prepared by the ion-exchange chromatography on SE-Sephadex yields a pattern which is identical with that of the authentic viral protease (Fig. 7). The same results were obtained with samples of the recombinant protease prepared by the two other isolation procedures. Likewise, the SDS-PAGE pattern of all samples of the recombinant p15$^{\underline{gag}}$ protease was

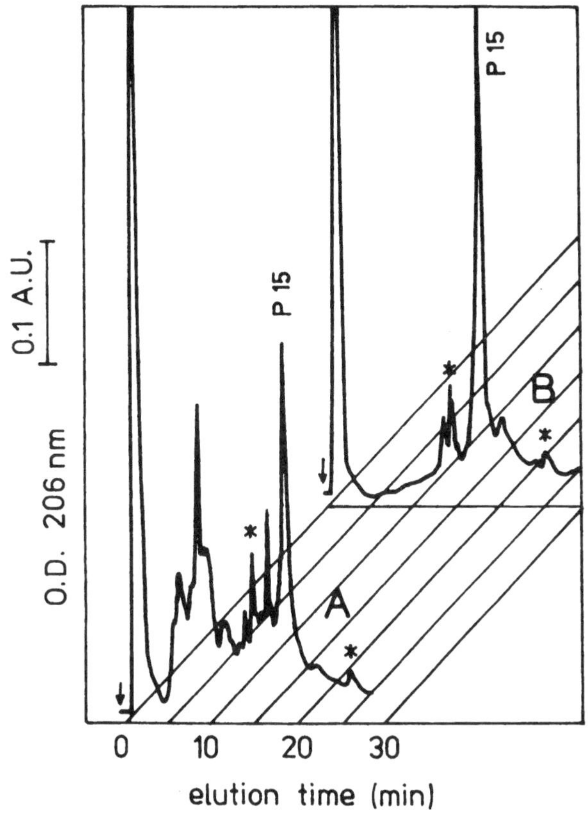

Fig. 6 Reverse-Phase HPLC on C-3 Ultrapore Column. Elution gradient 0 - 50% acetonitrile/25 min. A - chromatography profile of the recombinant p15<u>gag</u> protease, B - authentic p15<u>gag</u> protease

the same and identical to that of the authentic viral protease.

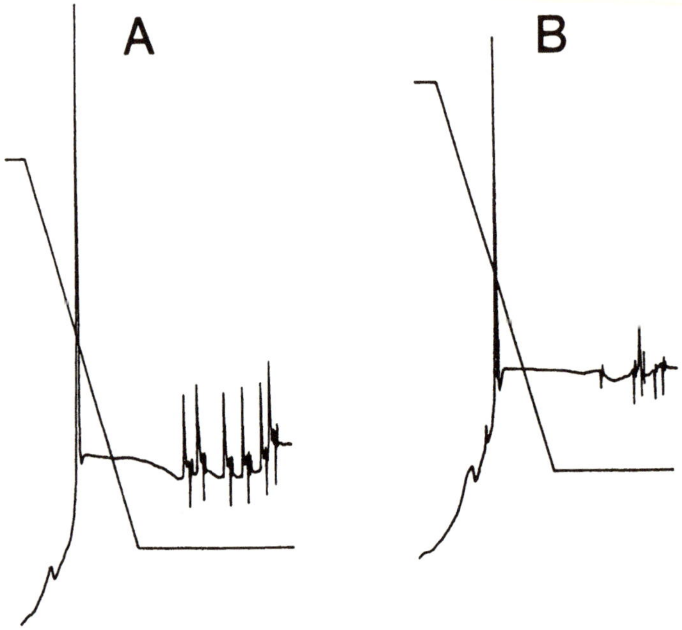

Fig. 7 FPLC of A) recombinant p15gag protease, B) authentic p15gag protease on a Mono S column. Elution gradient of 0 - 1.2 M NaCl in 0.05 M acetate buffer, pH 5, containing 0.05% mercaptoethanol, 0.005% Triton X-100.

Isoelectric point

The isoelectric point of the p15gag protease, determined by isoelectric focusing, was the same (pH 6.7) as that of the authentic viral protease.

Amino acid composition and N-terminal sequence

The amino acid composition of the recombinant protease is shown in Table 1. There are no essential differences from the composition of the authentic sample.

Table 1
Amino Acid Composition of Authentic p15$^{\underline{gag}}$ and Recombinant p15$^{\underline{gag}}$ Protease
Determined according to Spackman et al. (6).

Authentic protease (5)	Recombinant protease
Lys_3 His_3 Arg_{11} Asp_{12}	$Lys_{3,5}$ $His_{2,8}$ $Arg_{10,4}$ $Asp_{12,2}$
Thr_5 Ser_7 Glu_9 Pro_8	$Thr_{5,6}$ $Ser_{7,5}$ $Glu_{10,0}$ $Pro_{7,6}$
Gly_{12} Ala_8 Cys_1 Val_9	$Gly_{11,7}$ $Ala_{7,8}$ $Cys_{1,4}$ $Val_{9,4}$
Met_6 Ile_{11} Leu_{15} Tyr_1	$Met_{5,7}$ $Ile_{9,7}$ $Leu_{14,9}$ $Tyr_{1,2}$
Phe_1 Trp_1	$Phe_{1,7}$ $Trp_{0,9}$

The N-terminal amino acid sequence of individual samples of the recombinant protease differs with the starting material. The protease isolated from the cytoplasmic fraction showed the presence of N-terminal sequence Ala-Met-Thr-Met-Glu-His, which is by one residue, i.e. by the N-terminal leucine shorter than the sequence of the authentic viral p15$^{\underline{gag}}$ protease. The "shorter" protease was the only product isolated from the cytoplasm. Since the quantity of the protease increased with prolonged incubation periods the E. coli strain used obviously contains a strong amino-peptidase activity immediately cleaving all molecules of the protease not caught up in the inclusions.

Samples of the p15 protease isolated from the cytoplasmic fraction did not show proteolytic activity when assayed with denatured HSA as substrate. No activity was observed when the refolded (by 30-min treatment in urea followed by dialysis) substrate was used, either. These findings may indicate the importance of N-terminal leucine for the activity of the enzyme.

In contrast, when the inclusion bodies served as starting material, the recombinant protease was isolated in correctly processed form only, regardless of the procedure used for chromatography of the supernatants. These preparations were fully active and all showed N-terminal sequence Leu-Ala-Met-Thr-Glu-His, identical to the sequence of the authentic protease.

Acknowledgment

The skillful technical assistance of Ms Šenfelderová in the experimental part of this study is greatly acknowledged.

References

1. Trávníček, M., J. Říman. 1980. Biochem. Biophys. Res. Commun. 96, 1768-1777.

2. Sedláček, J., F. Kaprálek, P. Štrop, I. Pichová, V. Kostka, M. Trávníček. 1989. This volume, pp. 49-59.

3. Sedláček, J., M. Fábry, P. Kašpar, S. Zadražil, F. Kaprálek. 1986. Biotechnology & Bioindustry (Sofia) 6, 17-19.

4. Miller, M., J. Leis, A. Wlodawer. 1988. J. Mol. Biol. 204, 211-212.

5. Sauer, R.T., D.W. Allen, H.D. Niall. 1981. Biochemistry 20, 3784-3791.

6. Spackman, D.H., W.H. Stein, S. Moore. 1958. Anal. Chem. 30, 1190-1206.

BIOCHEMICAL CHARACTERIZATION OF HIV-1 PROTEASE PURIFIED FROM A BACTERIAL EXPRESSION SYSTEM

Thomas D. Meek[1], Brian D. Dayton[1], Brian W. Metcalf[1], Michael L. Moore[2], Joselina Gorniak[3], Martin Rosenberg[4] Christine Debouck[4], and James E. Strickler[3]

Departments of [1]Medicinal Chemistry, [2]Peptide Chemistry, [3]Macromolecular Sciences and [4]Molecular Genetics, Smith Kline & French Laboratories, King of Prussia, PA 19406.

Introduction

Within retroviruses, the *gag* and *pol* genes encode, respectively, the structural *gag* proteins of the virion and the enzymes essential for viral replication -- reverse transcriptase, ribonuclease H, and endonuclease. In the human immunodeficiency virus (HIV-1), the *gag* gene is expressed as a 55-kDa polyprotein precursor ($Pr55^{gag}$), while the translation of *pol* requires a -1 frameshift at the 3' end of *gag* to produce a 160-kDa fusion precursor, $Pr160^{gag\text{-}pol}$ (1-4). In all retroviruses, these *gag* and *gag-pol* precursors are processed into their respective mature protein products by the action of specific, virally-encoded proteases (5).

Retroviral proteases effect discrete digestion of their polyprotein substrates, as the cleavage sequences recognized by these proteases, including e.g. Ser/Thr-Xaa-Yaa-Tyr/Phe*Pro-Zaa (scissile bonds denoted by *) (6), are scarce within these polyproteins. This hexapeptide consensus sequence delimiting substrate specificity is unusual for the known endopeptidases, suggesting that the retroviral proteases have evolved for a specific function essential to the viral life cycle for which the cell cannot provide a substitute protease. Indeed, inactivation of protease within retroviral genomes by deletion or point mutation renders the resulting retroviruses incompetent to advance infection, and the corresponding virion particles are characterized by immature morphology and a decrease in reverse transcriptase activity (7-8). These findings suggest that inhibition of these essential and specific retroviral proteases may provide a new therapeutic opportunity for the treatment of AIDS and other diseases which result from infection by retroviruses.

Proteases of Retroviruses

When the primary sequence of a protease is known, its assignment to one of the four known classes of proteases (serine, thiol, metallo, or aspartyl) can be inferred from identification of a peptide sequence within it that is homologous to a conserved active site peptide that characterizes one of these protease classes. While limited homology exists among those retroviral proteases that have been purified and sequenced to date (9-12), two conserved domains, Leu-Val/Leu-Asp-Thr/Ser-Gly (domain I) and Ile/Leu-Ile/Leu-Gly-Arg-Asp/Asn (domain II), are present within the primary sequences of these proteases (13). Similar sequences may be found within the aspartyl proteases cathepsin D, renin, and the pepsins. In fact, each of the two aspartyl residues that constitute the catalytic machinery at the active sites of these proteases occurs within the tripeptide Asp-Thr-Gly. The presence of these two conserved tripeptide sequences within two distinct domains of the aspartyl proteases suggests that the bilobal structure of these enzymes may have resulted from gene duplication. Interestingly the retroviral proteases contain only one such conserved Asp-Thr/Ser-Gly sequence and consist of less than half the size of the aspartyl proteases. Pearl and Taylor have proposed from computer modeling of primary structure that the retroviral proteases could structurally resemble the aspartyl proteases if they existed as dimers of identical subunits, with each of the monomers contributing an Asp-Thr/Ser-Gly sequence to an active site formed at an interface of the subunits (14). We sought to address these issues for the retroviral protease from HIV-1.

Results

HIV-1 protease was obtained by expression in *Escherichia coli* of a restriction fragment derived from the BH10 clone (1) of the HIV-1 provirus (15). By locating within the translated HIV-1 genome the conserved domains I and II of the retroviral proteases, the coding region of the HIV-1 protease was found entirely within the *pol* open reading frame. Moreover, each of the domains was flanked by a consensus cleavage site, Ser/Thr-Xaa-Yaa-Phe*Pro-Zaa, such that specific autoproteolysis at these sites would release a 99-amino acid protein (10.8 kDa) bearing protease activity (Figure 1).

Two restriction fragments from the *pol* coding region of the proviral clone were inserted into the bacterial expression vector pAS to yield the PRO3 and PRO4 constructs (15). Each of these

|protease→
SerPheAsnPhe*ProGlnIleThrLeuTrpGlnArgProLeuValThrIleLysIleGlyGlyGlnLeuLysGluAla

--------- I ----------
LeuLeuAspThrGlyAlaAspAspThrValLeuGluGluMetSerLeuProGlyArgTrpLysProLysMetIle

Bcl I
GlyGlyIleGlyGlyPheIleLysValArgGlnTyrAspGlnIleLeuIleGluIleCysGlyHisLysGluIleCysGly

--------- II ---------
HisLysAlaIleGlyThrValLeuValGlyProThrProValAsnIleIleGlyArgAsnLeuLeuAsnLeuLeuThr

|reverse transcriptase→
GlnIleGlyCysThrLeuAsnPhe*ProIleSerProIleGluThrValProVal

FIGURE 1. Primary sequence of the HIV-1 protease coding region. The translated sequence spans nucleotides 1820-2158 (195-533 of the *pol* coding region) of the BH10 clone of HTLV-III_B (1). The conserved domains and the consensus cleavage sites of the retroviral proteases are indicated by broken and solid lines, respectively. The amino terminus of the protease and reverse transcriptase are indicated.

constructs encoded both conserved domains of the protease coding region and the flanking proteolytic cleavage sites. Instead of expressing the expected protein products of 20 kDa and 25 kDa, respectively, expression of the PRO3 and PRO4 constructs resulted in the accumulation of an 11-kDa protein product, suggesting that the recombinant HIV-1 protease was capable of autoprocessing, presumably at the consensus cleavage sites. In order to rule out the formation of the 11-kDa protein by adventitious proteolysis from a bacterial protease, a mutant was constructed containing an insertion of four codons (PRO4-BX) at the Bcl I site in PRO4. Induction of PRO4-BX resulted in the accumulation of a 25-kDa protein in lieu of the 11-kDa protein, suggesting that the 11-kDa protein observed in the PRO4 construct was indeed due to autoprocessing of the HIV-1 protease.

In addition, the recombinant HIV-1 protease processed a recombinant form of $Pr55^{gag}$ (p55) *in vivo* (15). Expression of the p55 vector in *E. coli* resulted in the formation of a 50-kDa protein, containing approximately 75% of $Pr55^{gag}$. Co-transformation in *E. coli* of the distinct p55 and PRO4 vectors resulted in processing of the 50-kDa *gag* precursor into two proteins of 24 kDa and 17 kDa. The sizes of these protein fragments are consistent with specific cleavage at the p17-p24 and p24-p15 junctions of $Pr55^{gag}$. Moreover, the amino terminus of the 24-kDa protein was found to be identical to that of authentic p24 from HIV-1 (15), indicating that the recombinant HIV-1 protease properly processed a recombinant form of its natural substrate, Pr55^{gag}. Co-transformation of p55 with the PRO4-BX or the PRO4-BS (termination codon inserted at the Bcl I site) mutants resulted in no processing of p55.

While induction of the PRO4 vector provided a highly active protease, it was expressed at very low levels. As a result, a specific and sensitive assay was needed as an adjuvant to purification of the protease from bacterial crude extracts. For this purpose, we purified a recombinant p55 protein from bacterial extracts, and then covalently labelled the purified p55 with a fluorescent ligand dansyl chloride. Recombinant p55, arranged as H_2N-p17-p24-p15-COOH, comprises most of the p17 region expressed as a fusion product, all of p24, and about half of p15. Dansylated p55 (Dns-p55) migrated as a single fluorescent band of about 50 kDa upon SDS-PAGE. Separate treatment with trypsin, chymotrypsin, or pepsin resulted in extensive digestion of Dns-p55, indicating that Dns-p55 was indeed a competent and non-specific protease substrate. Dns-p55 was incubated with crude extracts from identically transformed and induced *E. coli* transfected with the pAS vector, either with or without the PRO4 insert. No significant cleavage of Dns-p55 was observed in the control pAS extract, while the PRO4 extract effected limited digestion of Dns-p55 into discrete 41-kDa, 32-kDa, 24-kDa and 21-kDa protein bands (identical to that of Figure 2, lanes 4-9). These protein products are consistent with cleavage at the p24-p15 junction (41 kDa), the p17-p24 junction (32 kDa), or at both junctions to form the p24*gag* (24 kDa) and the p17*gag* (21 kDa) proteins.

In Figure 2 the use of the Dns-p55 assay to reveal protease activity in fractions obtained from analytical gel filtration of PRO4 crude extracts is illustrated. SDS-PAGE analysis of Dns-p55 proteolytic fragments, as well as the peptidolytic assay described below, was used to follow enzyme activity during purification of HIV-1 protease from PRO4 crude extracts. The protease was purified from PRO4 crude extracts to >90% homogeneity in two successive steps using ammonium sulfate precipitation and size exclusion HPLC using a column of TSK-G2000 SW (16). From determination of the amino acid composition and from NH_2-terminal sequencing (16), the protease, an 11-kDa protein by SDS-PAGE, was shown to consist of the 99 amino acids that fall between the two consensus cleavage sites in the PRO4 construct (Figure 1). The autoprocessing of the protease from its 25-kDa precursor occurs by cleavage at these sequences. The purified HIV-1 protease catalyzed the proteolysis of both Dns-p55 and unlabelled p55 into protein species of similar sizes.

The purified protease also hydrolyzes oligopeptides bearing a sequence similar to the known p17-p24 junction in Pr55^{gag} (-Ser-Gln-Asn-Tyr-Pro-Ile-Val- (2)). For example, the protease efficiently hydrolyzes the heptapeptide Ac-Ser-Gln-Asn-Tyr-Pro-Val-Val-NH_2 between the Tyr and Pro residues as shown by HPLC analysis (17). Incubation of the heptapeptide under identical conditions with bacterial crude extracts containing an induced pAS plasmid that lacked the PRO4 insert resulted in no peptidolysis of the substrate. These results indicate that the observed peptidolysis is not due to contaminating bacterial proteases.

Analysis of the peptidolytic fragments of this and other oligopeptide substrates of varying length by reversed-phase HPLC suffices as an efficient, quantitative assay for the protease, and such assays were used to characterize its enzymological properties. The kinetic constants obtained from double-reciprocal plots of initial velocity vs. peptide concentration are shown in Table 1 (17).

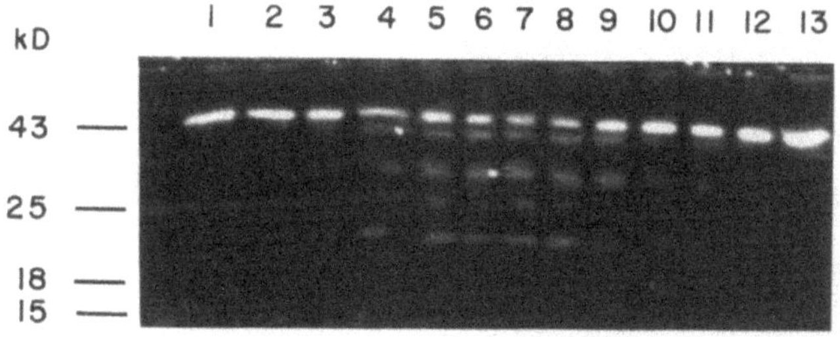

Figure 2. Assay of HIV-1 protease activity using Dns-p55 in fractions obtained from analytical gel filtration. PRO4 crude extract was chromatographed on a column of Sephacryl S-200 (Pharmacia, Superfine) equilibrated with 50 mM Tris-HCl (pH 7.0)-1 mM EDTA-0.1 M NaCl-1 mM DTT-0.15% Triton X-100 (TENDT buffer). Protease activity in 10-µL aliquots from each fraction was assayed by SDS-PAGE analysis of Dns-p55 digestion (0.45µg Dns-p55, 50 mM 2-(N-morpholino)ethanesulfonic acid (Mes; pH 6.0)-1 mM EDTA-0.2 M NaCl-1 mM DTT-0.1% Triton X-100 (MENDT buffer); 37° C), and proteins were visualized using a short-wavelength ultraviolet light source. Lane 1 contained no protease; lanes 2-13 are column fractions that containing the peak of protease activity.

Peptides 1 and 5 were not substrates for the protease. Of the oligopeptides tested that contained the consensus sequence Ser/Thr-Xaa-Yaa-Tyr/Phe*Pro-Zaa sequence, the hexapeptide Ac-Ser-Gln-Asn-Tyr-Pro-Val-NH_2 was the shortest oligopeptide with measurable activity as a substrate. Peptides 3 and 6 have been the substrates of choice for biochemical characterization of the protease.

We first sought to classify HIV-1 protease by its sensitivity to chemical inactivators which characterize the four known classes of proteases (18). Pre-treatment of the purified enzyme with high concentrations (0.1 mM, 100 mM) of compounds that inactivate or inhibit either the serine proteases (phenylmethylsulfonyl fluoride, N-tosyl-L-phenylalanine chloromethyl ketone, benzamidine) or the metalloproteases (1,10-phenanthroline, phosphoramidon, EDTA) had no

TABLE 1. Oligopeptide Substrates of HIV-1 Protease.

Peptide	K_m (mM)	V_{max} (rel.)	V_{max}/K_m (rel.)
1. AcSQNY*P-NH_2	---	0.00	0.00
2. AcSQNY*PV-NH_2	4.9±0.7	0.043	0.06
3. AcSQNY*PVV-NH_2	6.9±0.5	0.98	0.95
4. AcSQNY*PVVQN-NH_2	2.3±0.5	0.14	0.44
5. AcRASQNY*P-NH_2	---	0.00	0.00
6. AcRASQNY*PVV-NH_2	6.7±0.6	1.00	1.00

Reaction conditions: In 60 µL, 50 mM MES, pH 6.0, 1 mM EDTA, 1 mM dithiothreitol, 200 mM NaCl, 0.1% Triton X-100, 10% dimethyl sulfoxide, 0.5-20 mM peptide substrate, 30-150 ng purified HIV-1 protease; 37° C, 2-20 min.

effect on its activity. While leupeptin effected negligible inactivation of the protease, two other inactivators of the thiol proteases, N-ethyl maleimide and iodoacetamide, were potent inactivators, indicating that the enzyme possesses at least one sulfhydryl residue that is crucial for either catalytic or structural reasons. Three compounds that characterize the aspartyl proteases, Pepstatin A, N-diazoacetyl norleucine ethyl ester, and 1,2-epoxy-3-(4-nitrophenoxy)propane (EPNP), inactivated HIV-1 protease with IC_{50} values of 1.5 µM, 10 mM, and 450 µM, respectively. These results closely match results from similar inactivation studies conducted on the proteases from avian myeloblastosis virus (19) and bovine leukemia virus (20), and taken together with sequence homologies, suggests that the retroviral proteases belong to the class of aspartyl proteases.

In our hands, EPNP effected time-dependent, irreversible inactivation of HIV-1 protease (18). The action of EPNP on the protease appears to be active site-directed, in that the enzyme is protected from EPNP inactivation by the presence of a competitive inhibitor of peptidolytic activity. Upon characterization of this inactivation, the concentration of EPNP that caused half-maximal inactivation (K_{inact}) of the protease was 11 mM. As EPNP structurally resembles the P1 residue and scissile bond of the oligopeptide substrate (Figure 3), it is unsurprising that K_{inact} for EPNP is similar to the Michaelis constants of the substrates. EPNP has been shown to esterify one or both of the active site aspartyl residues of the pepsins (21-23). Hartsuck and Tang (22) have proposed that the inactivation of porcine pepsin by EPNP results from the nucleophilic attack of an unprotonated carboxylic acid group in the active site on the epoxide, in which the epoxide's

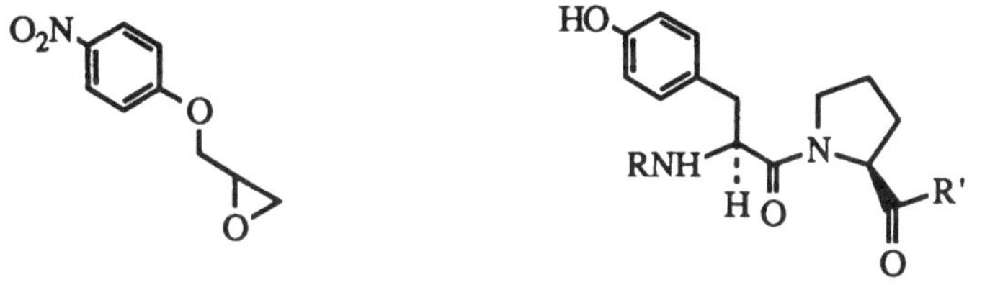

Figure 3. Structures of EPNP and the "Scissile Dipeptide" of an Oligopeptide Substrate.

oxygen is concomitantly protonated by an active site general acid. For the aspartyl proteases, such an inactivation mechanism could provide a plausible model for their proteolytic mechanisms in which each of the two conserved aspartyl residues assume opposite roles in general acid-general base chemistry.

The pH-dependence of EPNP inactivation of HIV-1 protease also implicates the esterification of an unprotonated carboxylic acid residue (18). Over a pH range of 3-8 (EPNP is chemically stable under these assay conditions), EPNP inactivation of the enzyme sharply decreases below pH 4, indicating that inactivation is abrogated following protonation of an enzymic group of pK= 3.8. Moreover, V_{max}, in a plot of log V_{max} vs. pH for the substrate AcRASQNYPVV-NH_2 also decreases as a residue of pK= 3.6 is protonated. Both of these results constitute compelling evidence that the active site of HIV-1 protease contains an unprotonated carboxylic residue that acts as a general base, and further supports the evidence afforded by protein sequence homologies that HIV-1 protease is indeed an aspartyl protease.

As described above, one curious aspect of the retroviral proteases is that the non-viral aspartyl proteases are at least twice as large and contain two active site aspartyl residues in the conserved sequence Asp-Thr-Gly. Pearl and Taylor have argued from computer modeling that the retroviral proteases could mimic aspartyl protease structure by dimerization of their monomeric subunits, such that each subunit contributes an Asp-Thr-Gly tripeptide sequence to an interfacial active site (14). We have investigated this intriguing possibility with the HIV-1 protease by comparing its subunit molecular weight of 11 kDa with the native molecular weight of the active enzyme determined by three methods: analytical gel filtration, density gradient centrifugation, and chemical crosslinking (18).

HIV-1 protease proteolytic (Figure 2) and peptidolytic activity eluted from a column of Sephacryl S-200 with an apparent molecular weight of 18-19 kDa, from which the Stokes radius was obtained (a= 2.0 nm). Similar chromatographic behavior has been reported for the protease from Moloney murine leukemia virus (24). From ultracentrifugation of the enzyme at pH 7.0 and 200 mM NaCl in a linear 17-35% glycerol gradient, active enzyme was found in a sharp band corresponding to a molecular weight of 21 kDa and a sedimentation coefficient of $s_{20,w} = 2.4 \times 10^{-13}$ sec. Siegel and Monty (25) have developed a protocol for calculation of native molecular weight from these two independently-determined parameters that minimizes inaccuracies esulting from protein asymmetry. According to their equation $M = 6\pi\eta N a s_{20,w}/(1-v\rho)$, in which η and ρ are the viscosity and density, respectively, of water at 20° C, N is Avogadro's number, and v is the partial specific volume of the protein (v= 0.757 cm^3/g; calculated from the amino acid composition (26)), substitution of the experimental values yields a value of $M = 22{,}000 \pm 3{,}000$. Thus, the active form of HIV-1 protease appears to be a dimer of identical 11-kDa subunits.

Additional evidence for this dimeric structure was obtained from chemical crosslinking studies. Treatment of the protease with the amine-specific crosslinking agent dimethylsuberimidate (27) at pH 8.5 revealed an immunoreactive 22-kDa protein band upon Western analysis. The ratio of 22-kDa and 11-kDa protein species over a 100-fold concentration range of the enzyme appeared invariant, suggesting that the crosslinking was occurring between two tightly associated subunits and not between free, unassociated monomers.

In summary, our preliminary characterization of a recombinant form of the HIV-1 protease has demonstrated that the enzyme's kinetic behavior is that of an aspartyl protease, and that its unique dimeric structure may be similar to the bilobal structure of this class of proteases. While we do not yet know that the active site is composed of residues contributed from each subunit, this idea is appealing in that the retroviruses would have "selected" the one class of protease that has structural symmetry at its active site, and, limited by the extreme brevity of their genomes, the retroviruses would have created such a protein by dimerization of a single, small gene product. It also remains to be seen whether or not the autoprocessing of the protease from the $Pr160^{gag\text{-}pol}$ requires dimerization of the protease domains within the polyproteins to initiate proteolysis upon a third polyprotein strand. In such an event, the association of protease domains within two polyprotein precursors may be regulated by changing cellular environment, such that the initiation of protein processing in the virion would occur only during virion assembly.

Acknowledgements

We thank Sam Franklin and Tom Smith for a gift of purified p55. Supported in part (B.D.D.) by a National Institutes of Health Cooperative Drug Discovery grant (AI 24845-02).

References

1. Ratner, L., Haseltine, W., Patarca, R., Livak, K.J., Starcich, B., Josephs, S.F., Doran, E.R., Rafalski, J.A., Whitehorn, E.A., Baumeister, K., Ivanoff, L., Petteway, S.R., Jr., Pearson, M.L., Lautenberger, J.A., Papas, T.S., Ghrayeb, J., Chang, N.T., Gallo, R.C., and Wong-Staal, F. 1985. Nature (London) **313**, 277-284.

2. Sanchez-Pescador, R., Power, M. D., Barr, P. J., Steimer, K. S., Stempien, M. M., Brown-Shimer, S. L., Gee, W. W., Renard, A., Randolph, A., Levy, J. A., Dina, D., and Luciw, P. A. 1985. Science **227**, 484-492.

3. Wain-Hobson, S., Sonigo, P., Danos, O., Cole, S., and Alizon, M. 1982. Cell **40**, 9-17.

4. Jacks, T., Power, M. D., Masiarz, F. R., Luciw, P. A., Barr, P. J. and Varmus, H. E. 1988. Nature (London) **331**, 280-283.

5. Dickson, C., Eisenman, R., Fan, H., Hunter, E., and Teich, N. 1982. In: RNA Tumor Viruses: Molecular Biology of Tumor Viruses, vol. 1 (R. Weiss, N. Teich, H. Varmus, and J. Coffin, eds.). Cold Spring Harbor Laboratory, Cold Spring Harbor, NY, pp. 513-648.

6. Pearl, L.H. and Taylor, W.R. 1987. Nature (London) **328**, 482.

7. Katoh, I., Yoshinaka, Y., Rein, A., Shibuya, M., Odaka, T., and Oroszlan, S. 1985. Virology **145**, 280-292.

8. Crawford, S. and Goff, S. P. 1985. J. Virol. **53**, 899-907.

9. Yoshinaka, Y., Katoh, I., Copeland, T.D., and Oroszlan, S. 1985. Proc. Natl. Acad. Sci. USA **82**, 1618-1622.

10. Sauer, R.T., Allen, D.W., and Niall, H.D. 1981. Biochemistry **20**, 3784-3791.

11. Yoshinaka, Y., Katoh, I., Copeland, T.D., Smythers, G.W., and Oroszlan, S. 1986. J. Virol. **57**, 826-832.

12. Yoshinaka, Y., Katoh, I., Copeland, T. D., and Oroszlan, S. 1985. J. Virol. **55**, 870-873.

13. Toh, H., Ono, M., Saigo, K., and Miyata, T. 1985. Nature (London) **315**, 691.

14. Pearl, L.H. and Taylor, W.R. 1987. Nature (London) **329**, 351-354.

15. Debouck, C., Gorniak, J. G., Strickler, J. E., Meek, T. D., Metcalf, B. W., and Rosenberg, M. 1987. Proc. Natl. Acad. Sci. USA **84**, 8903-8906.

16. Strickler, J. et al., unpublished results.

17. Moore, M. L., Bryan, W. M., Fakhoury, S. A., Magaard, V. W., Huffman, W. F., Dayton, B. D., Meek, T. D., Hyland, L., Metcalf, B. W., Strickler, J. E., Gorniak, J. G., and Debouck, C., unpublished results.

18. Meek, T. D., Dayton, B. D., Metcalf, B. W., Dreyer, G. B., Strickler, J. E., Gorniak, J. G., Rosenberg, M., Moore. M.L., Magaard, V. W., and Debouck, C. 1989., Proc. Natl. Acad. Sci. USA **86**, in press.

19. Dittmar, K.E.J. and Moelling, K. 1978. J. Virol. **28**, 106-118.

20. Katoh, I., Yasunaga, T., Ikawa, Y., and Yoshinaka, Y. 1987. Nature (London) **329**, 654-656.

21. Tang, J. 1971. J. Biol. Chem. **246**, 4510-4517.

22. Hartsuck, J.A. and Tang, J. 1972. J. Biol. Chem. **247**, 2575-2580.

23. James, M.N.G., Hsu, I.N., and Delbaere, L.T.J. 1977. Nature (London) **267**, 808-813.

24. Yoshinaka, Y. and Luftig, R.B. 1980. J. Gen. Virol. **48**, 329-340.

25. Siegel, L.M. and Monty, K.J. 1966. Biochim. Biophys. Acta 112, 346-362.

26. Cohn, E.J. and Edsall, J.T. 1943. In: Proteins, Amino Acids, and Peptides As Ions and Dipolar Ions, (Cohn, E.J. and Edsall, J.T., eds.). Hafner, New York, pp 370-381.

27. Davies, G.E. and Stark, G.R. 1970. Proc. Natl. Acad. Sci. USA **66**, 651-656.

EXPRESSION AND CHARACTERIZATION OF HUMAN IMMUNODEFICIENCY VIRUS-1 PROTEASE

M.C. Graves, J.J. Lim, M.A. Zicopoulos*, T.J. Stoller, M.C. Miedel*, Y.-C. E. Pan*, W. Danho*, and C.M. Nalin*

Department of Molecular Genetics and Department of Biological Chemistry*, Hoffmann-La Roche Inc., Nutley, NJ 07110

Introduction

The structural gag and enzymatic pol precursor proteins of retroviruses are processed by a viral proteinase (PR) to generate mature proteins (1). In the case of human immunodeficiency virus (HIV), a specific mutation in the PR domain results in the production of immature, non-infectious viral particles (2). Recent data demonstrate that the PR domain is autocatalytically released from the gag-pol fusion precursor to yield an 11 kDa active species (3). The mature p11 PR is capable of processing at least some of the specific sites in gag and gag-pol. Here we describe our progress on p11 PR expression in _E. coli_, purification, and characterization of the enzymatic activity.

Methods

Strains and plasmids

E. coli MC1061 was employed for general use, and the two PR expression plasmids, pPTΔN and pPT3 (3), were regulated by the lambda pL promoter and _c_I repressor. In pPTΔN, the gene encodes only the 11 kDa mature PR. Cells were co-transformed with one of two plasmids which contain different alleles of the _c_I repressor to allow derepression by either heat inactivation at 42°C (_c_Its, ref. 4) or chemical induction with nalidixic acid or mitomycin C (_c_I_ind_^{s}, ref. 5). The gene for mature PR also was placed under the control of the _lac_ operator (pPR5). In this case the _lac_ I^q repressor was encoded in the pDMI.1 plasmid

(6), and induction of PR was accomplished by addition of isopropyl thio-galactopyranoside (IPTG). In these systems, p11 PR accumulates in both the soluble and insoluble fractions of cell lysates and the proportion varies as a function of growth temperature, medium, etc. For large scale purification, cells were grown in 10 l fermenters in rich medium at the indicated temperature, induced for 4 hr, harvested, and stored at -80°C.

Standard TLC assay

Reaction volumes of 10 µl contained 100 mM Na phosphate, pH 6.0, 1 mM dithiothreitol (DTT), 0.5 µg iodinated peptide A (5-10 x 10^4 cpm) and the indicated amount of PR. PR was purified from the insoluble material as described below unless otherwise noted. Samples were incubated at 37°C for the indicated time and heated at 100°C for 10 min. Two µl aliquots were spotted on either cellulose (EM Science) or silica (Kodak) plastic-backed 20 x 20 cm TLC plates and subjected to 600 v•hr in pyridine/acetic acid/acetone/water [1:2:8:40, vol/vol, ref (7,8)]. Plates were dried, and the substrate and product were detected by autoradiography, excised, and quantitated by a ^{125}I gamma radiation counter.

Preparation of HIV PR

E. coli cells containing plasmid pPTΔN were induced at 42°C. Thawed cells were suspended (4 ml/g cell paste) in 50 mM Na phosphate, pH 6.5, 50 mM NaCl, 10 µg/ml DNase, 0.5 mM phenylmethylsulfonyl fluoride (PMSF), 0.5 mM EDTA, 0.5 mM DTT, disrupted by homogenization and subjected to centrifugation at 30,000xg for 15 min at 4°C. The pellet was suspended in Buffer A (50 mM Na acetate, pH 6.0, 0.5 mM PMSF, 0.5 mM EDTA, 0.5 mM DTT) supplemented to 7M guanidine HCl. The clarified supernatant was filtered through a 30,000 MW cut-off membrane and the effluent was diluted 40- to 60-fold with Buffer A. It was chromatographed on sulfopropyl (SP)-silica in Buffer A and eluted by a 0-1 M NaCl gradient. The PR preparation was used directly or further chromatographed on size exclusion resins. Samples were assayed for activity in the TLC- or HPLC-based assay and for PR protein by SDS-PAGE and immunoblot analysis.

Preparation of *E. coli* lysate

PR activity present in the soluble fraction of *E. coli* lysates was also characterized. MC1061 cells containing plasmids pPR5 and pDMI.1 were treated with 2 mM IPTG and grown at 25°C. Thawed cells were resuspended in 4 ml/gm of buffer containing 20 mM Na acetate, pH 5.2, 0.5 mM EDTA, 1 mM DTT supplemented with 1 μg/ml DNase and PMSF. The cell suspensions were sonicated with 5 30-second bursts and subjected to centrifugation at 30,000 x g for 45 min at 4°C. The protein concentration of the clear supernatant was usually 20-30 mg/ml. About 40 μg of protein was used for assay purposes. PR negative cells contained either parental plasmid or plasmid carrying another gene.

Peptide synthesis and analysis

Peptides were prepared as previously described (7). High specific activity iodinated peptides (2×10^7 cpm/μg) were prepared with the Pierce Iodo-bead procedure. Determination of the amino acid composition of substrates and products was as previously described (9).

Amino terminal sequencing of p11 PR

Bacterial cultures containing pPTΔN or pPT3 were prepared (3), sonicated, and the insoluble fraction resuspended in Laemmli buffer. Samples were subjected to SDS-PAGE and electrophoretically transferred to a polyvinylidene difluoride (PVDF) membrane (10). The membrane was stained with Coomassie Blue and the p11 PR band excised. Sequence analysis was performed using an Applied Biosystem (ABI) gas phase sequencer Model 470A (11). PTH amino acid derivatives were identified on-line with an ABI model 120A PTH Analyzer.

Results

The HIV gag and gag-pol precursor proteins are processed at numerous locations as depicted in Fig. 1. The sequences around the scissile bonds

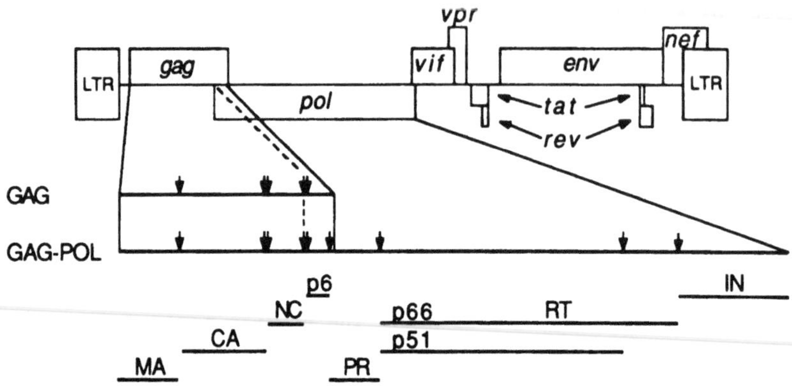

Fig. 1. HIV gag and pol precursor processing. The upper portion depicts the known coding regions of the HIV genome. The gag protein and gag-pol fusion protein are denoted by the heavy lines. Known processing sites are indicated by arrows. The resulting mature protein domains are shown at the bottom using the suggested standard nomenclature (13).

were deduced from the amino sequence analysis of purified viral proteins (13). While there is no homology common to all sites, the various sites have been grouped into three classes based on limited internal homology (14). The first class demonstrates the greatest degree of internal homology and is characterized by a Phe-Pro cleavage bond. Mature PR is the only domain in gag-pol that is bounded by "Phe-Pro" sites.

Earlier, we demonstrated that the mature p11 PR, expressed independently in E. coli, could correctly process the different sites in the pol precursor (3). The expression plasmid (pPTΔN, Fig. 2) that produced p11 PR encoded one extra Met prior to the Pro present at the mature p11 amino terminus. One of our first concerns was to ensure that the p11 PR amino terminus resulting from autoprocessing in E. coli (pPT3, Fig. 2) or synthesized directly (pPTΔN) was identical to that from the virus. p11 PR isolated from E. coli expressing pPT3 or pPTΔN was subjected to sequential Edman degradation. The amino terminus was Pro in both cases. These results indicate that first, autoprocessing of precursor is independent of background (E. coli or a viral particle), and that second, the initiating Met is removed from p11 PR in the case of pPTΔN expression. Since a stop codon is encoded following the PR domain in these plasmids, E. coli-derived p11 PR is identical in sequence to that from isolated virus.

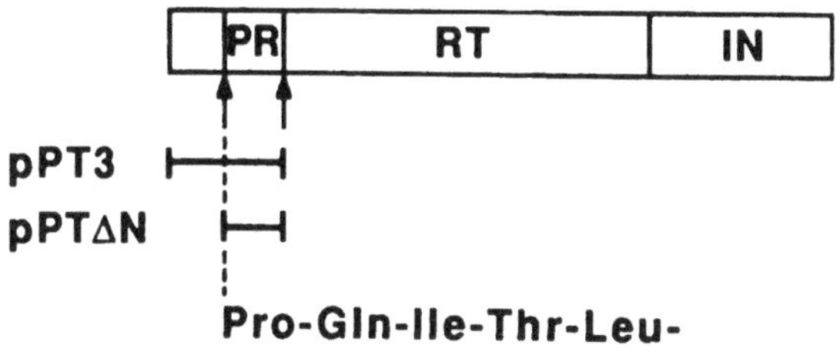

Fig. 2. Coding region of the HIV pol gene. Arrows delineate the portion that encodes the mature form of PR. Pol gene sections that are subcloned in pPT3 and pPTΔN are shown. The predicted (and actual) amino terminal sequence of p11 PR is shown.

PR assay development

Several assays were devised to aid PR purification and characterization. For purification purposes a rapid, sensitive assay was developed which employs a dodecapeptide similar to one of the known cleavage sites:

peptide A: H-Tyr-Val-Ser-Gln-Asn-Phe-Pro-Ile-Val-Gln-Asn-Arg-OH

Peptide A resembles the cleavage site that results in the CA p24 amino terminus, except that Phe was substituted for the normal Tyr. The amino terminal Tyr was added to allow radioiodination, which greatly increases the sensitivity of the assay. Substrate and products were routinely separated by electrophoretic thin layer chromatography and identified by autoradiography. This assay permitted the analysis of up to 18 samples per TLC plate in as little as 4 hours. As shown in Fig. 3 (lanes a and b), labelled substrate A and the N-terminal product can be easily separated. The assay was suitable for crude or purified enzyme preparations, although high salt and some detergents affected migration. As shown

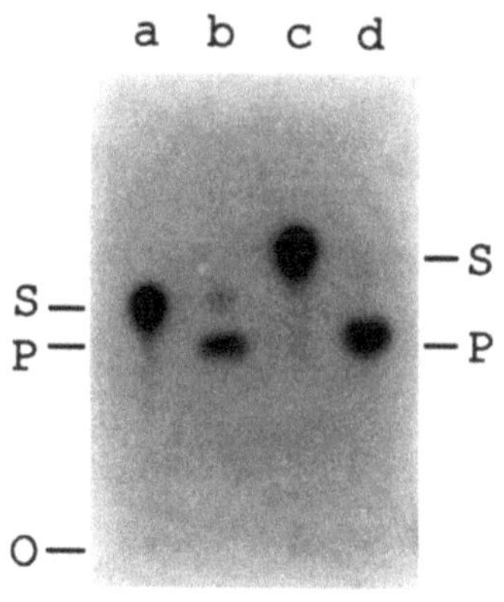

Fig. 3. Autoradiograph of cellulose TLC analysis of PR digestion products. O, origin; S, substrate; P, product. Lanes a and b contain peptide A, lanes c and d contain peptide B. PR enzyme was added to samples in lanes b and d as described in Methods.

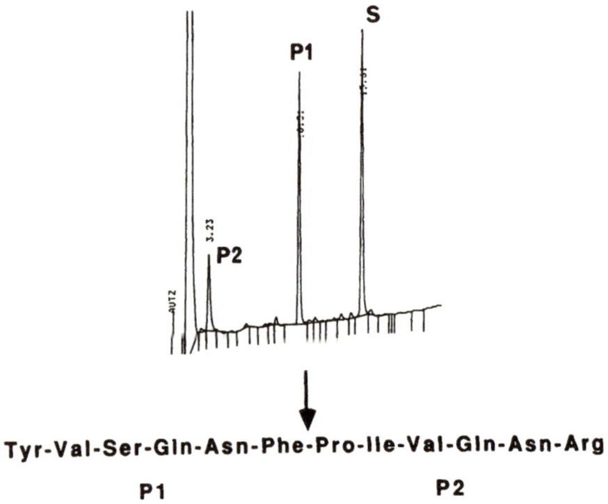

Fig. 4. HLPC analysis of PR digestion products. S, peptide A substrate; P, products. The 10 μl reaction contained 100 mM Na phosphate, pH 6.0, 10% glycerol, and 4.2 μg (3.4 nmol) of unlabelled peptide A (37°C for 20 min). The enzyme used was from the soluble fraction and had been purified by SP-silica chromatography. A Waters Novapak C_{18} column was employed. Buffer was 0.1% F_3CCO_2H with a 15-40% acetonitrile gradient; absorbance at 215 nm was monitored. PR cleavage indicated by the arrow.

below, E. coli proteases did not interfere with this or any other assay. HPLC analysis of peptide products is also useful for product recovery or for examining different peptide substrates, although it is not as rapid or sensitive as the TLC assay with radiolabelled substrates. A typical HPLC separation pattern of peptide A and its two products is shown in Fig. 4. Product peaks were collected and analyzed by composition and Edman degradation. Cleavage occurred between Phe and Pro, demonstrating that mature PR accurately hydrolyzes a peptide substrate.

PR expression and partial purification

Depending on the E. coli expression system and growth conditions employed, mature p11 PR accumulates in either the soluble or insoluble material of crude lysates, and we have optimized accumulation in either fraction. One variable examined was the mechanism of transcription induction from a repressed promoter. The PR gene in pPTΔN is under the control of the lambda pL promoter, and depending on the cI repressor allele used, induction can be mediated by heat, nalidixic acid or mitomycin C. A comparison of the immunoreactive material produced in these different systems is shown in Fig. 5. Even after long induction periods, both heat (42°C) and mitomycin C induction resulted in much higher accumulation levels than nalidixic acid. In all cases, the majority of the PR is insoluble. The purification advantage gained by

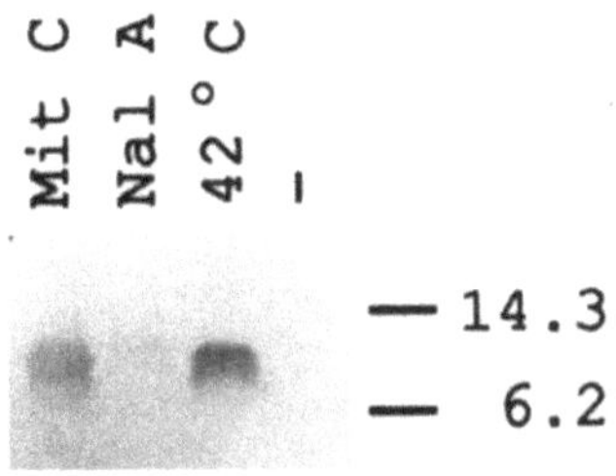

Fig. 5. Immunoblot analysis of E. coli expressing pPTΔN. Cells were lysed in Laemmli buffer, and 0.05 OD_{550} units of cells were analyzed by 15% SDS-PAGE. Proteins were transferred to nitrocellulose and probed with PR antiserum II (3). Induction was by 10 μg/ml mitomycin C or 40 μg/ml nalidixic acid for 5 hr, or by 42°C for 2 hr. Control (-) lane, uninduced cells.

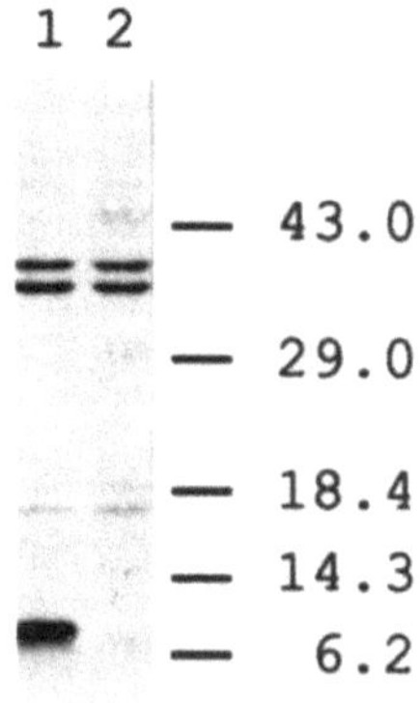

Fig. 6. Coomassie-stained SDS-gel of HIV PR expressed in E. coli. The pellet fraction from a cleared sonicate was resuspended in Laemmli buffer, and 0.1 OD_{550} of cells was analyzed by 15% SDS-PAGE. Lane 1, cells containing pPTΔN plasmid; lane 2, cells containing only repressor plasmid. Both strains were treated with 10 μg/ml mitomycin C for 3 hours. Molecular weights of markers are given in kDa.

starting with the insoluble fraction of crude E. coli lysates is apparent in Fig. 6, where p11 PR represents one of three major proteins. PR from such a preparation can be solubilized by 7M guanidine hydrochloride, refolded to an active species, and purified as outlined in Methods.

We also have optimized the accumulation of PR in the soluble fraction, mainly by the reduction of fermentation temperature to 25°C (see Methods). The PR present in crude lysates was assayed as above and demonstrated authentic peptide hydrolysis (Fig. 7). Activity is linear for up to 60 min at 37°C and then only falls off due to loss of substrate, not to a loss of intrinsic activity at that temperature. Note that even in crude lysates lacking HIV PR, no significant cleavage of the peptide substrate is evident, again demonstrating the absence of interference from E. coli proteases. In all assays, the characteristics of the soluble, non-denatured PR are the same as those of refolded PR.

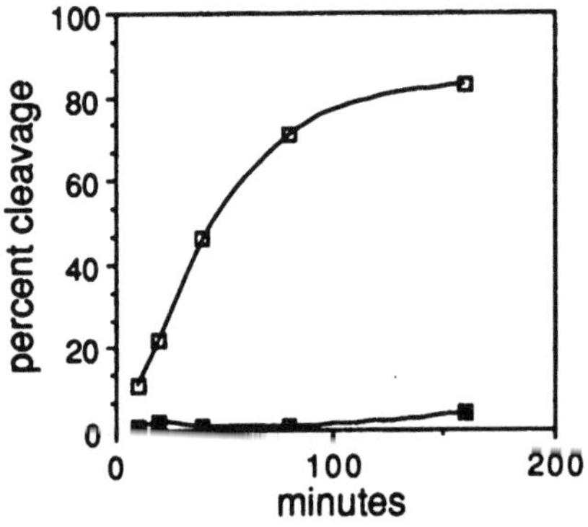

Fig. 7. Soluble HIV PR activity in E. coli lysates. Lysates were prepared from cells containing pPR5 (open symbols) or a plasmid lacking the PR gene (closed symbols). PR present in the soluble fraction was assayed by TLC analysis under standard conditions.

Characterization of enzymatic activity

We have found that the HIV PR accurately processes a peptide containing a "Phe-Pro" type of cleavage site (see above). Gag and pol precursors contain many processing sites that do not fall into this category, however. While our earlier work demonstrated the ability of p11 PR to process most, if not all, of these sites in the viral precursors, definitive evidence of peptide hydrolysis is also informative. A decapeptide analogous to the region spanning the amino terminus of gag NC p7 protein was synthesized with the expected cleavage between the two Mets.

peptide B: H-Tyr-Thr-Ala-Thr-Ile-Met-Met-Gln-Arg-Gly-OH

The peptide was iodinated and assayed under standard conditions. As shown in Fig. 5 (lanes c and d), treatment of peptide B with HIV PR converted the substrate to the expected single labelled product. HPLC analysis of unlabelled peptide substrate after addition of PR yielded two products (not shown). Amino acid composition of the products demonstrated that cleavage indeed occurred between the two Mets. These results offer conclusive proof that HIV PR can process sites other than "Phe-Pro".

PR enzymatic activity is stable in both crude and purified preparations. When preparations of 0.5 to 1 mg/ml total protein are stored at 4°C, activity losses are only a few percent per week. No loss has been observed over a 6 month period when the enzyme has been stored at -80°C. When PR was rapidly frozen, lyophilized to dryness, reconstituted, and

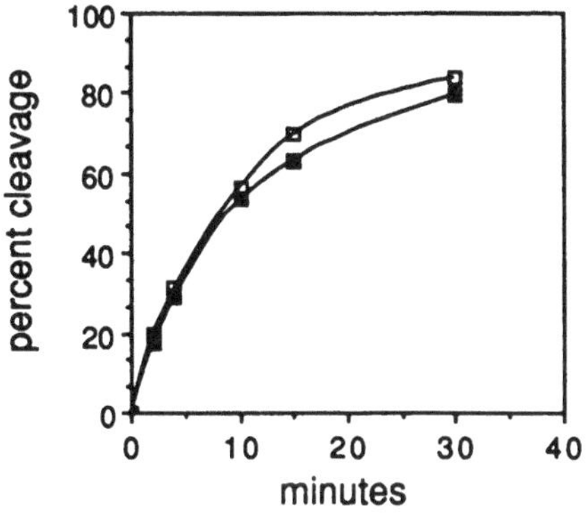

Fig. 8. Stability of HIV PR to freezing and lyophilizing. PR was incubated with peptide substrate for the indicated time under the standard assay conditions. Open symbol, PR stored at 4°C; closed symbol, lyophilized, and reconstituted.

assayed, no loss of activity was evident (Fig. 7). Activity is optimal between pH 5.5 and 6.0 under the conditions described and is stabilized by glycerol (10%, not shown).

Conclusions

The 11 kDa form of HIV PR expressed in E. coli is enzymatically active. Either by direct expression or by autoprocessing from a larger precursor, an 11 kDa species is generated that contains an amino-terminal Pro, and its amino acid sequence is probably identical to that produced in the virus. These results underscore the usefulness of this expression system when production from the natural source (i.e. virus) is undesirable. The bacterially-produced HIV PR authentically processes peptides containing viral precursor cleavage sites, including one containing a Met-Met scissile bond. The enzyme is stable to denaturation and refolding, lyophilization, and extended periods at 4°C or higher. By these and other criteria, we conclude that information gained from the characterization and inhibition of bacterially-derived PR enzyme will be the same as that for the enzyme present in viral particles.

Acknowledgements

The authors gratefully acknowledge the help of: L. Brink, R. Crowl, M. Kotler, R. Kramer, R. Makofske, L. Nieves, J. Switstok.

References

1. Dickson, C., R. Eisenman, H. Fan, E. Hunter and N. Teich. 1984. In: RNA Tumor Viruses, Vol. 1 (R. Weiss, N. Teich, H.Varmus, and J. Coffin, eds.). Cold Spring Harbor Lab., Cold Spring Harbor, New York, 2nd Ed., pp. 513-648.

2. Kohl, N.E., E.A. Emini, W.A. Schleif, L.J. Davis, J.C. Heimbach, R.A.F. Dixon, E.M. Scolnick, and I.S. Sigal. 1988. Proc. Natl. Acad. Sci. (USA) 85, 4686-4690.

3. Graves, M.C., J.J. Lim, E.P. Heimer, and R.A. Kramer. 1988. Proc. Natl. Acad. Sci (USA) 85, 2449-2453.

4. Bernard, H.-U. and D.R. Helinski. 1979. Methods Enzymol. 68, 482-492.

5. Crowl, R.M., R.P. Boyce, and H. Echols. 1981. J. Mol. Biol. 152. 915-819.

6. Certa, U., W. Bannwarth, D. Stuber, R. Gentz, M. Lanzer, S.F.J. LeGrice, F. Guillot, I. Wendler, G. Hunsmann, H. Bujard, and J. Mous. 1986. EMBO J. 5, 3051-3056.

7. Kotler, M., R.A. Katz, W. Danho, J. Leis, and A.M. Skalka. 1988. Proc. Natl. Acad. Sci. (USA) 85, 4185-4189.

8. Chen, R. 1976. Hoppe-Seyler's Z. Physiol. Chem. 357, 873-886.

9. Pan, Y.-C. E. and S. Stein. 1986. In: Methods of Protein Microcharacterization (J.E. Shively, ed.). Humana Press, Clifton, NJ. pp. 105-119.

10. Matsudaira, P. 1987. J. Biol. Chem. 262, 10035-10038.

11. Hewick, R.M., M.W. Hunkapiller, L.E. Hood, and W.J.Dreyer. 1981. J. Biol. Chem. 256, 7990-7997.

12. Leis, J., D. Baltimore, J.M. Bishop, J. Coffin, E. Fleissner, S.P. Goff, S. Oroszlan, H. Robinson, A.M. Skalka, H.M. Temin, and V. Vogt. 1988. J. Virol. 62, 1808-1809.

13. Henderson, L.E., T.D. Copeland, R.C. Sowder, A.M. Schultz, and S. Oroszlan. 1987. In: UCLA Symposium on Molecular and Cellular Biology, Vol. 71 (D. Bolognesi, ed.). Liss, New York. pp. 135-147.

14. Henderson, L.E., R.E. Benveniste, R. Sowder, T.D. Copeland, A.M. Schultz, and S. Oroszlan. 1988. J. Virol. 62, 2587-2595.

ANALYSIS OF HIV-1 PROTEASE VIA DELETION AND *IN VITRO* MUTAGENESIS STUDIES

S. Le Grice, A. Leuthardt, S. Reutener, R. Zehnle, J. Mous

Central Research Units, F. Hoffmann-La Roche & Co. Ltd., CH-4002 Basel, Switzerland

J. Brenner, J. Mills

Department of Biology, Roche Products Ltd., Welwyn Garden City, England

Introduction

In heterologous systems, the HIV-1 *pol* polyprotein is correctly processed to protease, reverse transcriptase, and endonuclease (1-6). Using a high level *E.coli* expression system (7), we have studied various aspects of protease-mediated processing. Elimination of the *pol* protease domain yielded an enzymatically inert reverse transcriptase/endonuclease polyprotein. Supplying protease *in trans* restored the processing pathway, showing that all cleavage events on the *pol* polyprotein are mediated by its own protease. We next altered one amino acid in the protease active site, and determined the effect in the context of a protease/reverse transcriptase and full length polyprotein. In both cases, enzymatically inactive polyproteins arose, and the processing programme was restored by supplying protease *in trans*. A similar result could be achieved with a genetically-engineered HIV-1 *gag*-protease fusion protein.

Proteases of Retroviruses

Results

Cleavage properties of a protease-deficient polyprotein

In preliminary experiments, we determined the effect of elimination of the pol protease domain on processing. Figure 1 illustrates that expression of the full-length pol polyprotein from plasmid pRT/CG (8) was followed by rapid cleavage to 64 and 52 kDa reverse transcriptase, 34 kDa endonuclease and the 10 kDa form of protease derived from the autocatalytic cleavage of the 17 kDa precursor (6,9,10). In contrast, *E.coli* harbouring the protease-deficient plasmid pRT/ENDO synthesised a 98 kDa polypeptide (Figure 1, middle). Cross-reactivity of this polypeptide with both HIV-1 positive serum and polyclonal endonuclease antibodies confirmed it as an uncleaved polypeptide containing reverse transcriptase and endonuclease. These results furthermore demonstrate that the internal reverse transcriptase cleavage site, as well as the site delineating reverse transcriptase from endonuclease, are both subject to cleavage by the pol-coded protease.

Secondly, we cloned the pol protease moiety as a separate transcriptional unit into plasmid pRT/ENDO, generating the plasmid pRT/ENDO-PROT, and supplied protease *in trans*. Under such circumstances, processing of the reverse transcriptase/endonuclease polyprotein was restored (Figure 1, lower series), demonstrating unambiguously that processing of the HIV-1 pol polyprotein is controlled exclusively by its protease moiety.

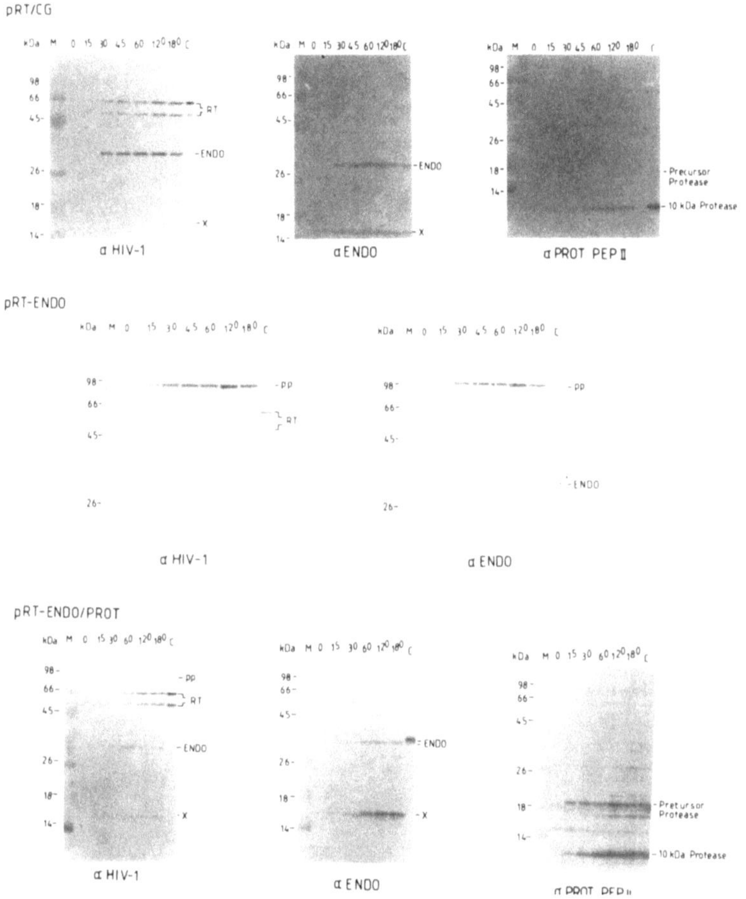

<u>**Figure 1**</u> Processing of full-length (pRT/CG, upper), protease-deficient (pRT-ENDO, middle) and <u>trans</u>-complemented (pRT-ENDO/PROT, lower) <u>pol</u> polyproteins. In the immunoblots presented, notation under each panel represents the antibody used, i.e. HIV-1, HIV-1-positive sera; ENDO, endonuclease; PROT PEP II, protease peptide II (6). Post-induction times (min) at which samples were analysed are indicated above the panels. pp, precursor polyprotein; M, molecular weight markers; m, control, purified reverse transcriptase (RT), endonuclease (ENDO) or protease. Polypeptide X arises internally from the endonuclease open reading frame.

Analysis of polymerase-encoded polyproteins containing a mutation in the protease domain

Within the 10 kDa portion of HIV-1 protease recently shown to be the biologically active form (9,10), the amino acid sequence -Asp-Thr-Gly- implicates it as a member of the family of aspartyl proteinases (11,12). Consequently, we elected to mutate the aspartic acid residue of this tripeptide to alanine (mutation D→A) and determine its effect on processing of various pol polyproteins. The plasmid pPROT-RT lacks the pol endonuclease domain, but the 81 kDa protease/reverse transcriptase portion could still be cleaved to the respective components (Figure 2A). In contrast, cells containing plasmid $pPROT^{D \to A}$-RT yielded the full length molecule, illustrating that the single amino acid alteration in the protease domain generates an enzymatically inert protein. Finally, adding a copy of the wild type protease sequence in trans, generating plasmid $pPROT^{D \to A}$-RT/PROT, resulted in release of the 64 and 52 kDa reverse transcriptase polypeptides, showing that the mutated protease/reverse transcriptase molecule had sufficient conformational integrity to be recognised by the trans complementing copy of protease.

Figure 2B shows a similar series of experiments performed on the full length pol polyprotein. Polyprotein containing the protease D→A mutation ($pPOL^{D \to A}$) accumulated as a 115 kDa molecule. Introduction of a second copy of wild type protease in trans ($pPOL^{D \to A}$/PROT) once more restored the polyprotein processing programme. The results with our D→A mutants confirm those in the previous section, illustrating that the pol encoded protease alone is responsible for all cleavage events on the polyprotein.

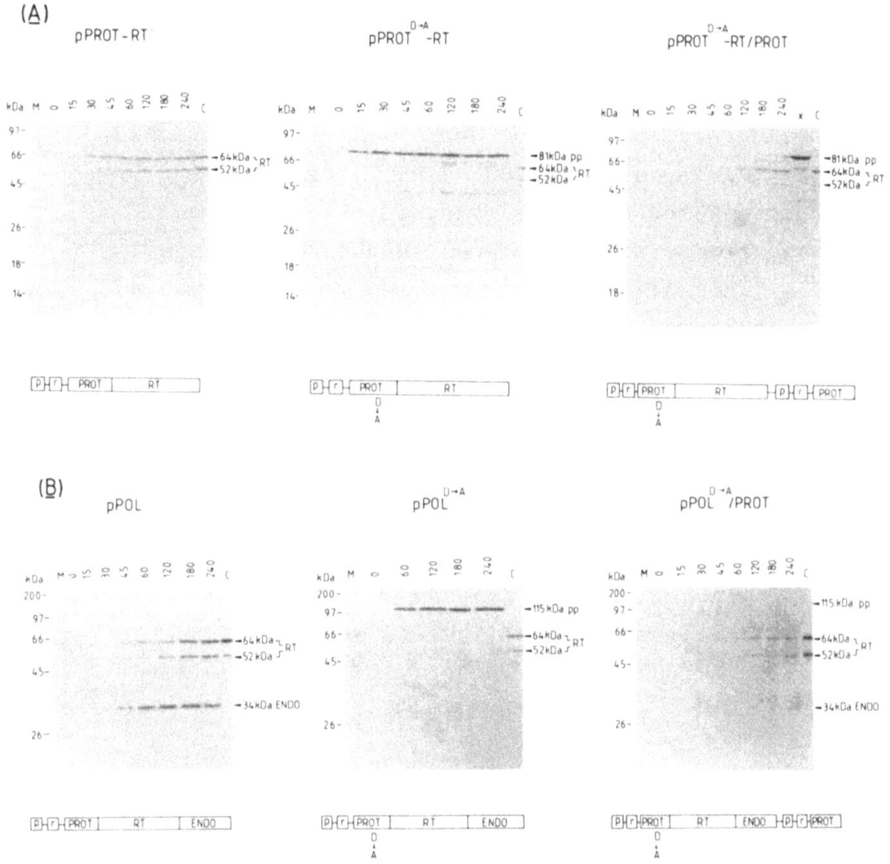

<u>Figure 2</u> (<u>A</u>) Processing of a wild-type (pPROT-RT), mutant (pPROT$^{D\rightarrow A}$-RT) and <u>trans</u>-complemented (pPROT$^{D\rightarrow A}$-RT/PROT) protease/reverse transcriptase polypeptide. All immunoblots were probed with a pool of HIV-1 positive sera (13). A brief description of the relevant expression cassettes is given below each panel. Within these cassettes, p is a controllable promoter (7) and r, a ribosome binding site. (<u>B</u>) Processing of wild-type (pPOL), mutant (pPOL$^{D\rightarrow A}$) and <u>trans</u>-complemented (pPOL$^{D\rightarrow A}$/PROT) forms of the full length HIV-1 <u>pol</u> polyprotein. Abbrevations are as before.

Enzymatic activity of pol-derived polyproteins

From our experiments, we generated three different polymerase-related polypeptides, i.e. protease-reverse transcriptase (plasmid pPROT$^{D \to A}$-RT), reverse transcriptase-endonuclease (plasmid pRT-ENDO) and protease-reverse transcriptase-endonuclease (plasmid pPOL$^{D \to A}$). This allowed the possibility of determining whether the reverse transcriptase component of these polypeptides displayed enzymatic activity. Hence, we purified and assayed all soluble polyprotein forms, the results of which are presented in Figure 3. From these results, it is clear that all polyprotein forms were inert, and that activity resided in preparations from either the wild-type or trans-complemented strains. From such an analysis, we conclude that the 64 and/or 52 kDa form of HIV-1 reverse transcriptase must be released from the polyprotein before enzymatic activity is established.

Analysis of a gag-protease fusion molecule

Although we have focussed primarily on the HIV-1 pol open reading frame, the major function of the pol-encoded protease is cleavage of the gag precursor (14). We therfore constructed the plasmid pGAG-PROT, wherein these two moeties were translationally fused. Under such conditions, we observed rapid release of the gag p24 component (Figure 4). When the D→A mutation was introduced into the protease portion (plasmid pGAG-PROT$^{D \to A}$), release of gag p24 was totally abolished, and could be restored by once more supplying a copy of wild type protease in trans (plasmid pGAG-PROT$^{D \to A}$/PROT), illustrating that the pol-encoded protease is also responsible for cleavage events releasing gag p24.

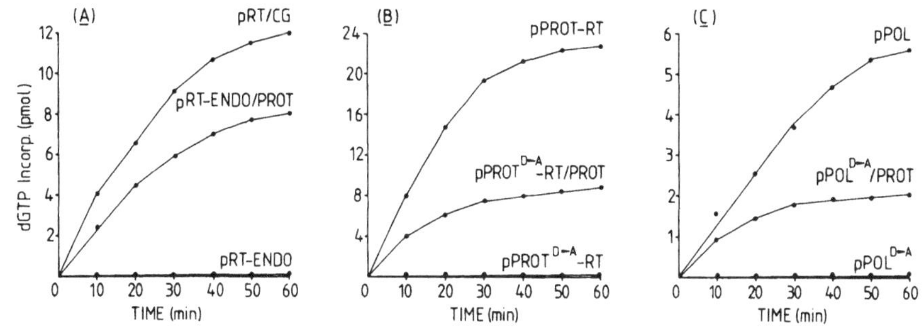

__Figure 3__ Analysis of reverse transcriptase activity in polyprotein forms. For each strain indicated, partial reverse transcriptase purifications and assays were performed according to published procedures (2). (A) Activity in a reverse transcriptase/endonuclease (pRT-ENDO) polypeptide, compared to that from a wild-type (pRT/CG) or __trans__-complemented (pRT-ENDO/PROT) strain. (B) Activity in a protease/reverse transcriptase polypeptide ($pPROT^{D\rightarrow A}$-RT), compared to that from a wild-type (pPROT-RT) or __trans__-complemented ($pPROT^{D\rightarrow A}$-RT/PROT) strain. (C) Activity in a full-length __pol__ polyprotein ($pPOL^{D\rightarrow A}$) compared to that of a wild type (pPOL) and __trans__-complemented strain ($pPOL^{D\rightarrow A}$/PROT). In all cases, the results presented are those of duplicate assays. In (A) - (C), the differences in incorporation reflect the amount of enzyme isolated rather than a difference in absolute activity. Although not shown here, we performed an immunological analysis on the proteins isolated from each strain using polyclonal antibodies against HIV-1 reverse transcriptase, establishing that correclty-sized reverse transcriptase or reverse transcriptase-containing polypeptides could be identified. This ruled out the possibility that a consequence of mutation was production of an insoluble protein.

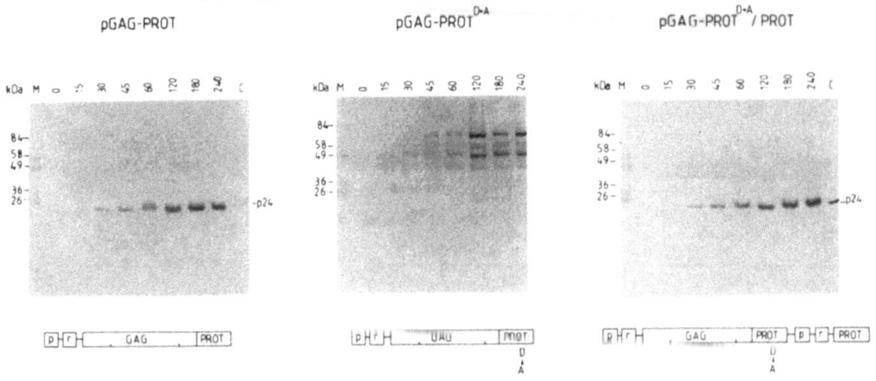

Figure 4 Processing of a wild-type (pGAG-PROT), mutant pGAG-PROT$^{D\to A}$) and trans-complemented (pGAG-PROT$^{D\to A}$/PROT) forms of a HIV-1 gag/protease fusion protein. Notations above and below each panel are as before. The gag cleavage sites leading to release of p24 are indicated on the expression cassettes. In the immunoblot presented here, antibody was a monoclonal antibody to HIV-1 gag p24.

Conclusions

We have shown here that the HIV-1 pol protease moiety is solely responsible for cleavage events on both the pol polyprotein and those leading to gag p24. In total, six cleavage events occur to release the gag- and pol-encoded polypeptides, i.e. (i) gag p17/p24, (ii) gag p24/p15, (iii) internal protease cleavage (iv) protease/reverse transcriptase (v) internal reverse transcriptase and (vi) reverse transcriptase/endonuclease. Comparison of amino acid sequences at these junctions (14) reveals no common recognition site, possibly suggesting that the environment in the vicinity of a cleavage site may play a role in its recognition. Additionally, we show here that the reverse transcriptase component of the pol polyprotein is only enzymatically active following its release from the polyprotein. This latter observation confirms earlier work of Witte & Baltimore on a murine retrovirus (15), who suggested it as a mechanism of preventing premature initiation of reverse transcription before viral infection.

Acknowledgements

We wish to thank Drs. M. Graves and E.P. Heimer, Hoffmann La-Roche, Nutley, N.J. for supplying antibodies to protease peptide II, and Drs. D.P. Bloxham and A.T. Kröhn, Roche Products Ltd. UK for providing logistical support.

References

1. Tanese, N., Sodroski, J., Haseltine, W.A., Goff, S.P. 1986. J. Virol. 59, 743.

2. Le Grice, S.F.J., Beuck, V., Mous,J. 1987. Gene 55, 95.

3. Farmerie, W.G., Loeb, D.H., Casavant, N.C., Hutchison III, C.A., Edgell, M.H. Swanstrom, R. 1987. Science 236, 305.

4. Hansen, J., Schultz, T., Moelling, K. 1987. J. Biol. Chem. 262 12393.

5. Larder, B., Purifoy, D., Powell, K., Darby, G. 1987. EMBO J. 6, 3133.

6. Mous, J., Heimer, E.P., Le Grice, S.F.J. 1988. J.Virol. 62, 1433.

7. Stuber, D., Ibrahimi, I., Cutler, D., Dobberstein, B. Bujard, H. 1984. EMBO J. 3, 3143.

8. Leuthardt, A., Le Grice, S.F.J. 1988. Gene (in press)

9. Debouck, C., Gorniak, J.G., Strickler, J.E., Meek, T.D., Metcalf, B.W., Rosenberg, M. 1987. Proc. Natl. Acad. Sci. (USA) 84, 8903.

10. Graves, M., Lim, J.J., Heimer, E.P., Kramer, R.A. 1988. Proc. Natl. Acad. Sci. (USA) 85, 2449.

11. Katoh., I., Yasunaga, T., Ikawa, Y., Yoshinaka, Y. 1987. Nature (London) 329, 654.

12. Pearl, L.H., Taylor, W.W. 1987 Nature (London) 329, 351.

13. Le Grice, S.F.J. Mills, J., Mous, J. 1988. EMBO J. (in press)

14. Ratner, L., Haseltine, W.A., Patarca, R., Livak, K.J., Starich, B., Josephs, S.F., Doran, E.R., Rafalski, J.A., Whitehorn, E.A., Baumeister, K., Ivanoff, L., Petteway, S.R., Pearson, M.L., Lautenberger, J.A., Papas, T.S., Ghrayeb, J., Chang, N.T., Gallo, R.C., Wong-Staal, F. 1985. Nature (London) 316, 277.

15. Witte, O.N., Baltimore, D. 1976. J. Virol. 26, 750.

IN VITRO MUTATION OF THE ASP-25 BLOCKS AND PEPSTATIN A INHIBITS THE ACTIVITY OF THE HUMAN IMMUNODEFICIENCY VIRUS (HIV) ENCODED PROTEASE.

K.von der Helm, S.Seelmeier, and H.Schmidt,

Max von Pettenkofer Institute, University of Munich, D-8000 Munich 2, FRG

ABSTRACT

The protease encoded by the human immunodeficiency virus (HIV) processes the viral gag and gag-pol protein precursor by post-translational cleavage. In this study we demonstrate by site-specific mutagenesis (asp -> thr) and by pepstatin A inhibition that the recombinant HIV protease is an aspartic-type protease. Furthermore incubation of HIV infected H9 cells with pepstatin A inhibited part of the intracellular processing of the HIV gag protein yet had no apparent toxicity on HIV infected cells during 48hrs incubation.

INTRODUCTION

Virus encoded proteases were first described in avian and murine retroviruses (1,2). They were shown to process the gag protein precursor to yield mature gag-proteins that are required for assembly of infectious virus particles (3,4,for review see 5). Retroviral proteases appear to be highly specific for their substrate (6-9).

It is conceivable that an inhibitor specifically directed against a retrovirus protease could block the processing of the viral gag proteins, thereby preventing the production of infectious virus without interfering with host cell (proteolytic) activities. Murine retrovirus with a defective

Proteases of Retroviruses

protease was shown to produce immature virus particles lacking infectivity (3,4).

Before searching for suitable inhibitors of the virally coded proteases the functional groups in the active site of the enzyme have to be revealed. Comparisons of the amino acid sequences of retroviral proteases show the presence of a conserved asp-thr-gly (asp-ser-gly) sequence similar to that of aspartic proteases (5,10,11) and a structural model for a viral aspartic protease has been recently proposed (11). It has also been reported that the proteolytic activities of bovine leukemia virus (BLV), Moloney murine leukemia virus (MoMLV) and human T-cell leukemia virus (HTLV I) could be inhibited in vitro by Pepstatin A, an aspartic protease inhibitor although at a relatively high drug concentration (12). Here we describe studies (13) on the inhibition in vitro and in part in vivo of the HIV protease confirming the suggestion (5,10,11) that the HIV coded protease is an aspartic-type protease and demonstrating that it can be inhibited by pepstatin A.

RESULTS

A site specific mutation of the HIV protease was made (14) on a recombinant pTZ (Pharmacia) construct (BglII-EcoRI fragment of the HIV DNA-clone (LAV) (15)) by replacing the aspartic acid residue in the conserved asp-thr-gly sequence with a threonine residue.(13). When the wild type construct (pTZprt+) was expressed in E.coli and the lysates incubated in vitro, immunoblot analysis showed the formation of processed viral reverse transcriptase molecules and no polypeptide of the predicted length of Mr 98 (13).

The mutationally altered construct (pTZprt-) had the unprocessed, primary translation product of Mr 98 indicating that the Asp -> Thr mutation blocked the virus specific proteolytic activity which normally processes the prt-pol recombinant protein . The 98kd recombinant protein of the mutant (pTZprt-) could, however, be processed *in trans* to the reverse transcriptase when wild type protease (pTZprt+) was incubated with the mutant enzyme, (13) indicating that the asp -> thr mutation did not render the protein uncleavable but blocked the active enzymatic site.

The proteolytic activity of the mutant was also examined towards its specific substrate, pr 55gag. Recombinant wild type (pTZprt+) or the mutant protease (pTZprt-) were incubated with purified and Triton X-100 disrupted HIV, which normally still contain a significant amount of unprocessed pr 55gag. The mutant protease failed completely to cleave the pr 55gag, whereas the wild type enzyme processed the gag protein precursor (13). Using a recombinant pr55 gag protein (a (SacI-HindII) fragment of the HIV DNA (15) comprising the entire gag protein sequence) as substrate, pr 55gag was processed to intermediate precursor (ipr40), p24 and to a protein migrating similar as the viral p17. The mutant protease did not cleave the recombinant gag precursor at all (13).

As these results indicated the enzyme to be an aspartic protease we used the in vitro system of recombinant viral protease and gag precursor to test the effect of inhibitors of aspartic or other types of proteases (13). The proteolytic activity was significantly inhibited by pepstatin A (16) an aspartic protease inhibitor. At a concentration of 5×10^{-5} M a partial inhibition is indicated; at 2.5×10^{-4} M 50% inhibition (IC50) appeared to be achieved, and at 1 mM pepstatin A the inhibition seemed to be complete. The processing of pr 55gag was also but less markedly inhibited by PCMB. PCMB affects sulfhydryl groups, but is not selective for cysteine proteases; specific cysteine inhibitors as cystatin (19) and E-64 did not inhibit. The serine protease inhibitor PMSF did not affect the protease. Two more inhibitors (VdLPFFVdL (17); PHPFHL°VIHK (18)) both effectively inhibiting other aspartic acid proteases (17,18) did not inhibit the HIV protease at a concentration of 10^{-3} M.

In a preliminary experiment the effect of pepstatin A was tested on the processing of the gag precursor in HIV infected cells. The experiment was carried out by dissolving pepstatin, which is poorly water soluble, in

DMSO, and diluting it to the culture medium of non-infected or HIV-infected H9 cells.(13). After incubation for 48h, cell viability determined by trypanblue staining of uninfected and infected cells was not significantly altered by pepstatin A treatment as compared to the not-treated control cells. The intracellular processing of HIV specific gag processing was, however, affected by pepstatin A: the processing of gag precursor, pr 55gag, to the intermediate precursor pri40 was markedly decreased analysed by immunoblot analysis. The amount of released virus particles was reduced after pepstatin A treatment of culture cells as compared to non-treated cells judged by measuring the amount of HIV antigen after aliquots of ultracentrifugation pelleted culture medium were immunoblotted.

DISCUSSION

We confirmed and extended (13) the assumption that the HIV coded (gag processing) protease is an aspartic or aspartic-like protease by the following results: (i) site directed mutagenesis of the aspartic acid residue in the presumed active site (conserved asp-thr-gly sequence) to a threonine residue abolished the proteolytic activity of the recombinant HIV protease, (ii) an inhibitor of aspartic proteases clearly affected in vitro the activity of the recombinant protease.

The experiments were done with recombinant protease protein expressed in E.coli similar to that reported recently (20-23). In preceding experiments we found it extremely difficult to prepare native HIV protease, probably due to the minute amounts present in virus particles and probably because of its potential hydrophobic character.

The recombinant wild type protease processed native precursor protein pr 55gag prepared from HIV material and precursor protein pr 55gag produced by recombinant technique.The cleavage activity could be blocked by the asp->thr mutation of the recombinant protease. Recent reports on in vitro (24) and in vivo (25) mutations confirm this.

The results with the site specific mutagenesis also confirmed earlier suggestions (21) that the recombinant HIV encoded protease is responsible for cutting the reverse transcriptase from a HIV POL gene product, a prt and RT precursor; they corroborate indirectly the autocatalytic processing of the protease (22). This has interesting implications for future therapeutic considerations, because blocking the protease activity at the level of the gag-pol precursor stage probably abolishes both protease and reverse transcriptase activity.

An inhibition in vitro of the protease activity was achieved by pepstatin A (13). Some inhibitory effects of pepstatin A on the avian retroviral protease p15 in vitro (6) and on focus formation of murine sarcoma virus (MSV) in vivo (26) had been reported before, and very recently also on HIV protease (27).

The experiment with pepstatin A in cell cultures was a very preliminary study (13) to see if pepstatin A showed any effect to intracellular HIV gag processing and whether or not it would be toxic to the cells. The results show that pepstatin A (i) is able to inhibit at least part of the intracellular HIV specific gag processing (pr 55gag -->ipr 40) and (ii) does not cause serious toxic side effects on the cells during the incubation time of 48 hrs.

Experiments are now in progress testing the infectivity of virus progeny released after exposure of HIV infected H9 cells to pepstatin A and related inhibitors.

The concentration of pepstatin A for 50% inhibition (IC50) of the HIV protease in the range of 10^{-5} M and 10^{-4} M is comparable to concentrations described by Katoh et al. (12) for inhibiting other retroviral proteases. The IC50 values for the inhibition of other aspartic proteases by pepstatin A are in the range of 10^{-6} - 10^{-8} M. Thus the inhibition of pepstatin A against the HIV protease is less efficient than against other aspartic proteases. However, pepstatin A is the only one of three aspartic acid inhibitors we tested (13) which does affect the HIV protease, whereas the other inhibitors effectively inhibiting aspartic proteases do not inhibit the HIV protease.

This and the fact that the viral protease has less than half the molecular weight of other aspartic proteases (11) implies that the two types of aspartic proteases must have a distinct molecular structure. This is an intriguing motivation to search for more specific inhibitors or possibly design new ones which will discriminate between viral proteases and their "counterparts" and affect preferentially the viral enzyme thus rendering HIV to a poorly or non-infectious virus.

ACKNOWLEDGEMENT

We thank F.Deinhardt for his interest in and support of this work, H.Fritz for very helpful discussions, L.Gürtler and J.Eberle for kindly providing cultures of HIV- and non- infected H9 cells, part of the HIV used in this study and HIV antiserum, S.Wain-Hobson for providing the HIV (BRU) DNA clone (15), and H.Wolf and F.Schwarzmann for synthesis of oligonucleotides. The work was supported by the Bundesministerium f.F&T.

LITERATURE

1 von der Helm,K. (1977) Proc.Natl.Acad.Sci.USA 74,911-915

2 Yoshinaka,Y. & Luftig,R.B. (1977) Cell 12,709-719

3 Crawford,S. & Goff,S.P. (1985) J.Virol. 53,899-907

4 Katoh,I., Yoshinaka,Y.,, Rein,A., Shibuya,M., Odaka,T. & Oroszlan,S. (1985) Virology 145,280-292

5 Kräusslich,H.G. & Wimmer,E. (1988) Ann.Rev.Biochem. 57:701-754

6 Dittmar,K.J. & Moelling,K. (1978) J.Virol. 28,106-118

7 Khan,A.S. & Stephenson,J.R. (1979) J.Virol. 29,649-656

8 Yoshinaka,Y., Katoh,I., Copeland,T.D. & Oroszlan,S. (1985) J.Virol. 55,870-873

9 Kräusslich,H.G. & von der Helm,K. (1987) Virology 156,246-252

10 Toh,H., Ono,M., Saigo,K. & Miyata T. (1985) Nature 315,691-692

11 Pearl,L.H. & Taylor,W.R. (1987) Nature 328,482

12 Katoh,I., Yasunaga,T., Ikawa,Y. & Yoshinaka,Y. (1987) Nature 329,654-656

13 Seelmeir,S., Schmidt,H., Turk,V. & von der Helm,K. (1988) Proc.Natl.Acad.Sci.USA 85,6612-6616

14 Morinaga,Y., Francheschini,T., Inouye,S., Inouye,M. (1984) Biotechnology, 636

15 Alizon,M., Sonigo,P., Barré-Sinoussi,F., Chermann,F.C., Tiollais,P., Montagnier, L. & Wain-Hobson,S. (1984) Nature 312,757-760

16 Umezawa,H., Aoyagi,T., Morishima,H., Matsuzaki,M., Hamada,M. & Takeuchi,T. (1970) J.Antibiot. 23,259-262

17 Pohl,J., Zaoral,M., Jindra,Jr.,& Kostka,V. (1984) Anal. Biochem. 139 265-272

18 Szelke,M., Leckie,B., Hallet,A., Jones,D.M., Sueiras,J., Atrash,B.,& Lever,A.F. (1982) Nature 299 555-558

19 Turk,V., Brzin,J., Longer,M., Ritonja,A., Eropkin,M., Borchart,U. & Machleidt,W. (1983) Hoppe-Seyler's Z.Physiol.Chem.364,1487-1496

20 Kramer,R.A., Schaber,M.D., Skalka,A.M., Ganguly,K., Wong-Staal.F. & Reddy.E.P. (1986) Science 231,1580-1584

21 Farmerie,W.G., Loeb,D.G., Casavant,N.C., Hutchison,C.A.I, Edgell,M.H. & Swanstrom,R. (1987) Science 236,305-308

22 Debouck,C., Gorniak,J.G., Strickler,J.E., Meek,T.D., Metcalf,B.W. & Rosenberg,M. (1987) Proc.Natl.Acad.Sci.USA 84,8903-8906

23 Graves,M.C., Lim,J.J., Heimer,E.P.,& Kramer,R. (1988) Proc.Natl.Acad.Sci.USA 85, 2449-2453

24 Mous,J., Heimer,E.P.,& LeGrice,St.F.J. (1988) J.Virol. 62, 1433-1436

25 Kohl,N., Emini,E.A., Schleif,W.A.,et.al. (1988) Proc.Natl.Acad.Sci.USA 85,4686-4690

26 Yuasa,Y., Shimojo,H., Aoyagi,T. & Umezawa,H. (1975) J.natn.Cancer Inst. 54,1255-1256

27 Hansen,J., Bilich,B., Schulze,Th., Sukrow,S.& Mölling,K. (1988) EMBO J.7, 1785-1791

MUTATIONAL ANALYSIS OF THE HIV-1 PROTEASE: IMPORTANCE FOR FUNCTION OF SEQUENCES WITHIN AND FLANKING THE PROTEASE DOMAIN.

D.D, Loeb, C.A. Hutchison, III

Department of Microbiology and Immunology, Curriculum in Genetics, Program in Molecular Biology and Biotechnology, University of North Carolina at Chapel Hill, Chapel Hill, NC 27599

L. Everitt, W.G. Farmerie, R. Swanstrom

Department of Biochemistry, and Lineberger Cancer Research Center, University of North Carolina at Chapel Hill, Chapel Hill, NC 27599

Introduction

Retroviruses encode their capsid proteins and the virus-specific enzymes in the viral *gag* and *pol* genes, respectively. Multiple proteins are encoded by each gene so that the mature gene products are generated only after post-translational proteolytic processing. These processing events are carried out by a viral protease (PR; encoded in *pol*), and processing is a required step in the generation of infectious virus (reviewed in ref. 1).

The *pol* gene product is initially expressed as a fusion protein that contains both *gag* and *pol*. In the human retrovirus HIV-1, the fusion of the upstream *gag* reading frame with the downstream *pol* reading frame occurs by an occasional translational frameshifting event (2).

We will review our efforts to understand proteolytic processing by the HIV-1 protease. This discussion will include an examination of the role of sequences immediately flanking the protease domain, and a consideration of important sequences within the protease domain that are required for function.

Proteases of Retroviruses

Are Sequences Upstream of the Protease Domain Needed For Protease Function?

The location of the frameshifting event has been identified (2). It occurs 66 codons before the *gag* termination codon at the third codon of the *pol* reading frame. The *gag* and *pol* reading frames overlap by approximately 68 codons.

Information about the role of specific sequences in the *gag/pol* overlap region that precedes the PR domain can be obtained by comparing the sequences in this region between two related viruses, HIV-1 (3) and HIV-2 (4).

```
A.
      ***                      **
HIV-1 FFREDLAFLQGKAREFSS EQTR ANSPTRRELQVWGRDNNSP
HIV-2 FFRTGPLGKEAPQLPRGPSSAGADTNSTPSGSSSGSTGEIYAAREKTE

                          *    *        **  ** **
HIV-1                  SFAGADRQGTV  SFNFPQVTLWQRPLVT
HIV-2 RAERETIQGSDRGLTAPRAGGDTIQGATNRGLAAPQFALWKRPV
                      /                 /

B.
       ***      * *     *      *****        *
HIV-1 NFLGKIWPSYKGRPGNFLQ SRPE PTAPPEESFRSGVETTTPP
HIV-2 GFLGLGPWGKKPRNFPVAQVPQGLTPTAPPVDPAVDLLEKYMQQGKRQR

                          *    *     *  * **** *
HIV-1                  QKQEPIDKELY  PLTSLRSLFGNDPSSQ
HIV-2 EQRERPYKEVTEDLLHLEQGETPYREPPTEDLLHLNSLFGKDQ
                      /                 /
```

Figure 1. Comparison of HIV-1 and HIV-2 amino acid sequences in the *pol* and *gag* reading frames within the *gag/pol* overlap region. The single letter amino acid code is used to compare the *pol* reading frame (panel A) and the *gag* reading frame (panel B) within the *gag/pol* overlap region. Alignments were maximized using equivalent amino acid substitution groups but only those positions showing sequence identity are indicated (*). Vertical arrows are used to show the position of the *gag/pol* frameshift (2) near the beginning of the sequence and the position of the beginning of the protease domain (5) near the end of the sequence. The horizontal arrows indicate the position of a direct repeat that appears in the HIV-2 sequence. The HIV-1 sequence is from Ratner et al. (3) and the HIV-2 sequence is from Guyader et al. (4).

Surprisingly, there is very little sequence similarity. There are three identical amino acids in the *pol* reading frame at the site of the HIV-1 frameshift (Phe Phe Arg; Fig. 1A; the equivalently positioned amino acids in the *gag* reading frame are also identical; Fig. 1B). There is also a stretch of five identical amino acids in the *gag* reading frame (Pro Thr Ala Pro Pro, starting at position 24 in the HIV-1 overlap region; Fig. 1B), and significant sequence similarity in *pol* after the start of the protease domain. Otherwise, very little similarity is apparent between HIV-1 and HIV-2 in either the *gag* or *pol* reading frames within the overlap. The length of the overlap is very different for HIV-1 and HIV-2, 205 base pairs in HIV-1 versus 282 base pairs in HIV-2. In some isolates of HIV-1, there is a 36 base pair duplication near the middle of the overlap region. HIV-2 shows evidence of a 51 base pair duplication just preceding the PR domain and including the equivalent sequence that represents the amino terminus of the HIV-1 PR (Fig. 1A).

This extreme variability is contrasted with the amino acids encoded in the protease domain which show 45% identity, 22% conservative substitutions, and no insertions or deletions in comparing HIV-1 and HIV-2. Similarly, the 100 codons in *gag* preceding the *gag/pol* overlap can be aligned between these two viruses to give 59 positions of identity and 16 positions of conservative substitutions with the introduction of one small deletion and one small insertion. This lack of sequence conservation in the *gag/pol* overlap region makes it unlikely that this region encodes amino acids in either the *gag* or *pol* reading frames that serve an important enzymatic or structural role for the virus. One possibility is that the overlap region provides a spacer between the *gag* and *pol* domains in the *gag/pol* precursor. Such an interpretation is consistent with the observation that most of this region can be deleted without affecting the processing of the *pol* domain (Fig. 2). An additional role for part of the overlap region may be to promote the frameshifting event between the *gag* and *pol* reading frames. Both HIV-1 and HIV-2 encode an inverted repeat in the primary nucleotide sequence just downstream of the HIV-1 frameshifting site. It has been suggested that the inverted repeat could form a hairpin that would stall ribosome movement and thus promote frameshifting (2).

The lack of sequence conservation in the *gag/pol* overlap region suggested that these sequences may not be important in generating an active protease. We tested this possibility by deleting the *gag/pol* overlap region and then measuring the ability of the deleted genome to express a functional protease. In these experiments we have used a bacterial expression system in which the HIV-1 *pol* gene is expressed in *E. coli* and the primary translation product is processed by the viral protease to give mature-sized products (6,7).

CLONE	DELETION	PR	PROC
ART1 E3	-25	+	+
ART1 E66	-16	+	+
ART1 E46	-10	+	+
ART1 E42	-7	+	±
ART1 E71	-4	+	±
ART1 E81	-1	-	-
ART1 E72	1	-	-
ART1 E86	3	-	±
BRT1 E16	5	-	-
BRT1 E3	6	-	-

Firgure 2. The effect of N-terminal deletions on the ability of PR to process. Deletions at the 5′ end of the *pol* gene were tested for their ability to generate a mature PR (p11) and for their ability to cleave at the other processing sites in the *pol* precursor when expressed in *E. coli*. The column labelled DELETION indicates the deletion point in the clone before the mature N-terminus, and positive numbers represent the residue within the mature protein at which the clone begins. All deletion clones have approximately 30 amino acid residues fused onto the amino terminus, which is contributed from the vector. The ability to generate a mature PR is indicated by a + or - in the PR column. A clone's ability to process the other *pol* derived proteins is indicated by either a +, ±, or - in the PROC column. A + indicates only mature products were seen, a - inidcates no mature products, and a ± indicates some mature products and some unprocessed products. All deletions beyond residue 6 were negative for PR and PROC.

A nested set of deletions of the *pol* reading frame was constructed and each was tested for protease activity (Fig. 2). Most of the *pol* reading frame upstream of the PR domain could be deleted without loss of processing. Deletions close to the processing site that generates the amino terminus of the protease reduced the efficiency of processing, but did not affect the accuracy of processing. This was true of deletions to within four amino acids of the processing site. However, deletions that extended even a few amino acids into the PR domain blocked normal processing. Thus we conclude that in the bacterial expression system most of the sequences in the *gag/pol* overlap region that precede the PR domain are not important for expressing a functional PR.

Are Sequences at the C-Terminus of the Protease Domain Needed for Function?

Graves et al. (8) have shown that expression of only the PR domain yields an active protein, and similar results have been obtained using a chemically synthesized protein (9,10). However, it is not known what fraction of the PR is active in these cases. The *pol* auto-processing event seen in bacteria indicates that either every protease molecule synthesized is active or that each molecule is folded properly to serve as substrate. Auto-processing can occur when all sequences downstream of the PR domain are deleted from the *pol* gene-expressing plasmid (8). To determine if C-terminal PR sequences are important for function, we introduced termination codons at various amino acids near the C-terminal end of the protease (as defined by protein sequence analysis; 5,11,12) and then assayed for the production of an N-terminal processed protease in bacteria. We found that the removal of a single amino acid from the C-terminus destroys protease activity. Thus, like the N-terminus, flanking sequences do not appear to play an important role in generating protease activity while the amino acids just within the N- and C-terminal processing sites are required for activity.

Domains Within the Protease That Are Important For Function

Mutagenesis can be used to identify functionally important amino acids within a protein. We have used this approach to study the HIV-1 protease. In our initial test of functionally important residues, we examined the sensitivity to mutation of a 13 amino acid domain (7) which contains homology to the active site of aspartic proteinase (13). We found this domain to be very sensitive to amino acid change, the majority of the substitutions destroyed activity (Fig. 3). The sensitivity of this domain was compared to the sensitivity of a flanking domain of 12 amino acids downstream; in this case a minority of substitutions affected activity (Fig. 3). Thus the sequence spanning the presumed active site is very sensitive to mutagenesis, an observation that is consistent with this domain playing an important role in protease function. Also, this observation provides strong genetic evidence that the HIV-1 protease is a member of the aspartic proteinase family of proteases.

e A L L D T G A D D T V L e E M s L P G r W k p k

Figure 3. Sensitivity of a 25 residue region to mutational inactivation by nonconservative substitutions. The sensitivity of a position to mutational inactivation is indicated by the case of the one letter amino acid code. An underlined uppercase residue indicates that all nonconservative substitutions tested inactivate PR function. An uppercase residue indicates that the most active nonconservative substitution is intermediate in its PR activity. A lowercase residue indicates that all nonconservative substitutions tested had no effect on PR activity. The above amino acid sequence represents residues 21 to 45 in the PR domain.

More recently we have extended this analysis to encompass the entire protease domain. We have found a total of three sensitive domains, the other two being a glycine-rich region in the center of the protease and a hydrophobic domain near the C-terminus. A complete description of the mutational analysis will be given elsewhere (Loeb et al., in prep.).

Acknowledgements

This work is supported by grants from the NIH to R.S. and C.A.H. R.S. also is supported by a Junior Faculty Award form the American Cancer Society.

References

1. Oroszlan, S. and R. Luftig. Processing of Retroviral Proteins. In: Current Topics in Microbiology and Immunology (P.K. Vogt and R. Swanstrom, eds.). (in press).

2. Jacks, T., M.D. Power, F.R. Masiarz, P.A. Luciw, P.J. Barr, and H.E. Varmus. 1988. Nature 331, 280-283.

3. Ratner, L., A. Fisher, L.L. Jagodzinski, H. Mitsuya, R.S. Liou, R.C. Gallo, and F. Wong-Staal. 1987. AIDS Research and Human Retroviruses 3, 57-69.

4. Guyader, M., M. Emerman, P. Sonigo, F. Clavel, L. Montagnier, and M. Alizon. 1987. Nature 326, 662-669.

5. Lillehoj, E.P., R.H.R. Salazar, R.J. Mervis, M.G. Raum, H.W. Chan, N. Ahmad, and S. Venkatesan. 1988. J. Virol. 62, 3053-3058.

6. Farmerie, W.G., D.D. Loeb, N.C. Casavant, C.A. Hutchison III, M.H. Edgell, and R. Swanstrom. 1987. Science 236, 305-308.

7. Loeb, D.D., C.A. Hutchison III, M.H. Edgell, W.G. Farmerie, and R. Swanstrom. 1989. J. Virol. (in press).

8. Graves, M.C., J.J. Lim, E.P. Heimer, and R. Kramer. 1988. Proc. Natl. Acad. Sci. USA 85, 2449-2453.

9. Schneider, J. and S.B.H. Kent. 1988. Cell 54, 363-368.

10. Nutt, R.F., S.F. Brady, P.L. Darke, T.M. Ciccarone, C.D. Colton, E.M. Nutt, J.A. Rodkey, C.D. Bennett, L.H. Waxman, I.S. Sigal, P.S. Anderson, and D.F. Veber. 1988. Proc. Natl. Acad. Sci. USA 85, 7129-7133.

11. Di Marzo Veronese, F., T.D. Copeland. A.L. DeVico, R. Rahman, S. Oroszlan, R.C. Gallo, and M.G. Sarngadharan. 1986. Science 231, 1289-1291.

12. Lightfoote, M.M., J.E. Coligan, T.M. Folks, A.S. Fauci, M.A. Martin, and S. Venkatesan. 1986. J. Virol. 60, 771-775.

13. Toh, H. R. Kikuno, H.Hayashida, T. Miyata, W. Kugimiya, S. Inouye, S. Yuki, and K. Saigo. 1985. EMBO J. 4, 1267-1272.

SYNTHESIS AND ACTIVITY OF HTLV-I PROTEASE AND ITS MUTEINS

M. Hatanaka, S.H. Nam
Institute for Virus Research, Kyoto University, Sakyo-ku, Kyoto 606, Japan

Introduction

We have molecularly cloned a novel full-length provirus genome of HTLV-I from virus producing MT-2 cells (1). This clone, designated as HTLV1C, harbors an open reading frame between the gag and the pol gene which corresponds to the protease encoding region of other retroviruses (2). This open reading frame has never been identified in the nucleotide sequence of other isolates (3, 4). We have reported that amino acid sequences strictly conserved in other retroviral proteases are also found in the possible protease-encoding sequence of HTLV1C. We used a vaccinia virus expression vector to synthesize the HTLV-I protease and its muteins, and investigated the biological function and expression mode of the protease gene.

Results

Processing of gag precursor polyprotein in recombinant vaccinia expression vector system

The 2kb RsaI-ApaI fragment harboring the entire gag and pro open reading frame (ORF) was purified and introduced into downstream of the promoter of p7.5K gene which expresses at both early and late time of the vaccinia virus infection cycle.

Proteases of Retroviruses

Plasmid p7.5gagN was introduced into CV-I cells infected with vaccinia virus by using the calcium phosphate coprecipitation technique. The transient expression of the gag gene products in CV-I cells was examined by Western blotting using anti-gag monoclonal antibodies. All the gag proteins transiently expressed from plasmid p7.5gagtN comigrated on SDS-polyacrylamide gel electrophoresis with those corresponding to HTLV-I gag proteins synthesized in HTLV-I producing MT-2 cells or to virion-associated gag proteins of HTLV-I.

Proteolytic activity resides within the putative gene product of proORF

To demonstrate that the processing of the gag precursor polyprotein observed in the expression system using the vaccinia virus vector is accomplished by HTLV-I protease encoded by nucleotide sequences in the regions of proORF, the protease-encoding sequences were deleted by available restriction enzyme digestions. Our results demonstrated that the region spanning nucleotide position 2228 to 2485 of proORF is important for the processing of Pr53gag into the mature gag products. By contrast, the downstream region from the OxaNI site of proORF has no relation to the proteolytic function of the protease.

Processing defect of proORF-deletion mutant is compensated by supplying active protease

To obtain evidence that the intact protease molecule directly cleaves the gag precursor polyprotein, we tried to determine whether the gag precursor polyprotein directed by the proORF-deletion mutant is able to be processed by the wild type protease in trans action. To elucidate this point, we tried to establish an enzyme-substrate reaction system in vivo. A DNA transfection experiment in the vaccinia virus expression vector system is suitable for this kind of study because the co-introduction of two independent recombinant DNA's can be achieved with maximal efficiency by using the recombinant vaccinia virus as the helper virus. As an efficient donor of the active form of protease a recombinant vaccinia virus expressing the protease was constructed by using hemagglutinin gene-mediated homologous recombination. The resultant recombinant vaccinia virus, WRgagN, harboring the gag and proORF, produced the processed gag products which were identical to those synthesized in MT-2 cells. In control experiments, recombinant vaccinia virus harboring the gagORF alone was constructed and designated as WRgag ΔTS. No correct processing has occurred in cells infected with WRgag ΔTS. We conclude that the HTLV-I protease encoded by the proORF is able to process the gag precursor polyprotein by trans action.

Importance of aspartic acid residues in HTLV-I protease

The alignment of amino acid sequences of the HTLV-I protease and of other retroviral proteases revealed two stretches of strictly conserved sequences, i.e. L-L-D-T-G and I-I-G-R-D (2, 5). We have shown above that the region containing these conserved amino acid sequences plays an important role in the expression of proteolytic activity.
The fact that sequence L-L-D-T-G is also found in aspartyl proteases such as pepsin implies that the aspartic acid residue located in this sequence might be essential for the catalytic action of the HTLV-I protease. To identify the role of the aspartic acid in position No 64 in the catalytic activity of HTLV-I protease, site specific mutagenesis was used to replace the aspartic acid residue by glycine. The

resultant mutant plasmid p7.5gagN AspI directed 53K gag precursor polyprotein, but no other specifically processed proteins; this suggests that the HTLV-I protease may belong to aspartyl proteases.

Possible model of biosynthesis of HTLV-I protease

To obtain evidence that the HTLV-I protease is synthesized by suppression of gag termination codon by frameshifting, we introduced mutation of six consecutive adenines (nucleotide positions 2064 through 2069) located in the overlapping region of the gag and proORF; the mutant plasmid obtained was designated as p7.5gag fs22. These consecutive adenine residues are expected to be involved in putative frameshifting in analogy to other retroviruses such as bovine leukemia virus (BLV). In mutant p7.5gag fs22, three A to T gag transversions were generated without disruption of the phase of the gag gene. Processing of gag precursor polyprotein was no longer observed indicating that the HTLV-I protease protein is synthesized via ribosomal frameshift at six consecutive adenine residues located in the 3' terminal region of gag gene.

We may speculate that the mode of generation of the mature protease is the following: a frameshift gives rise to a gag-pro fusion protein with protease activity, which is expected to be 76kDa in size. This gag-pro fusion protein is rapidly cleaved by its own or cellular protease to generate free protease molecules that can act in trans on 53K gag precursor polyprotein. To confirm the scheme described above, it is important to demonstrate the presence of the 76K precursor polyprotein and then to identify the protease involved in it. In HTLV-I producing cells, the 76K protein is, however, hardly detectable, probably because of the inefficient frameshifting and/or rapid processing. To mimic the natural frameshifting we introduced one base to adjust the coding frames of the gag and pro genes at the consecutive adenine residues located in the 3' terminal region of gag gene of the protease-negative or -positive plasmids (p7.5gagN AspI, p7.5gagN AspII and jp7.5gagN). As expected, the programmed size of products, i.e. of 68K and 65K proteins, was directed by p7.5gagfs19AspI OE and p7.5gagfs19AspII Kp, respectively. The results confirmed the view that the 76K protein observed is the real gag-pro precursor polyprotein.

Conclusions

The biological activity encoded by the putative protease gene (pro) of human T-cell leukemia virus type I (HTLV-I) was investigated using a vaccinia virus expression vector. The 53K gag precursor polyprotein was processed to the mature p19, p24 and p15 gag precursor proteins when the gag and protease encoding sequence were expressed under the control of the vaccinia virus promoter, suggesting that the protease may be synthesized by a mechanism of ribosomal frameshift. The processing defect of the mutants (muteins) could be compensated by co-introduction of a wild type construct into the cell. This demonstrates that the pro gene encodes biologically active protease molecules which are capable to process *in vivo* the gag precursor polyprotein *in trans*. A study of a variety of mutants (muteins) constructed *in vitro* revealed that the protease consists of a nonessential carboxy-terminal region and a part essential for its activity including the putative catalytic residue of aspartic acid. Further, a cluster of adenine residues positioned in the region overlapping the gag and pro gene was shown to be involved in ribosomal slipping which is responsible for the synthesis of the protease. To mimic the formation of 76K gag-pro precursor polyprotein formed by ribosomal slipping, the ribosomal slipping, the coding frames of the gag and pro gene were modified. The processing of the gag-pro precursor polyprotein occurred only when the intact protease gene was present, implying that the cis-acting function of HTLV-I protease is necessary to trigger the initial cleavage event for its own release from the precursor protein.

References

1. Mori, K., H. Sabe, H. Siomi, T. Iino, A. Tanaka, K. Takeuchi, K. Hirayoshi, M. Hatanaka. 1987. J.Gen.Virol. 68, 499.

2. Nam, S.H., M. Hatanaka. 1986. Biochem.Biophys.Res.Commun. 139, 129.

3. Ratner, L., S.F. Josephs, B. Starcich, B. Hahn, G.M. Shaw, R.C. Gallo, F. Wong-Staal. 1985. J.Virol. 54, 781.

4. Seiki, M., S. Hattori, Y. Hirayama, M. Yoshida. 1983. Proc.Natl.Acad.Sci.USA 80, 3618.

5. Katoh, I., T. Yasunaga, Y. Ikawa, Y. Yoshinaka. 1987. Nature(London) 329, 654.

ACTIVATION AND REGULATION OF AVIAN SARCOMA LEUKEMIA VIRUS (ASLV) PROTEASE

M. Kotler

Department of Molecular Genetics, The Hebrew University-Hadassah Medical School, Jerusalem, Israel

Introduction

In the last few years there has been an increased interest in retroviral proteases (PR's). The stimulation to work on the PR's was augmented by the finding that their inhibition prevents virus propagation (1), a finding which has been confirmed later with HIV-1 (2). It is therefore believed that the specific inhibitor(s) of the viral protease may be useful for therapeutic intervention in retrovirus-induced diseases, such as AIDS. Hence, most of the research in this field is directed toward biochemical characterization of the protease and to preparation of specific inhibitors of this enzyme. In this presentation I would like to concentrate mainly on the biological functions fulfilled by the protease in virus propagation and on mechanisms which control its activation. The information gained from these aspects of retrovirus replication is very important, not only from the general biological point of view, but also for designing suitable inhibitors of retrovirus replication.

Proteases of Retroviruses

Results

Expression of Retrovirus PR

The dimeric RNA genomes of retroviruses are encapsidated in a symmetric icosahedral core that is surrounded by a lipoprotein envelope (3). The viral genes, gag, pol and env are translated to produce precursor polypeptides which are cleaved after virion assembly. The infectious virus contains processed "mature" proteins (3). Cleavage of the gag and pol precursors of avian sarcoma leukemia viruses (ASLV) (4-8), murine leukemia viruses (MuLV) (1, 9-11), and human immunodeficiency virus (HIV-1) (12) requires the activity of a viral protease (PR) which is itself encoded in one or more of the precursors. ASLV PR is a 15 kD (124 amino acids) protein and is synthesized as part of two large precursor proteins which are translated from the genomic size mRNA. The first of the two precursor peptides is $Pr76^{gag}$, which includes PR at its C-terminus, and the second one is $Pr180^{gag-pol}$ which contains the gag and pol sequences (Fig. 1). The translation of this large protein occurs via a frameshift at the termination codon for PR (13).

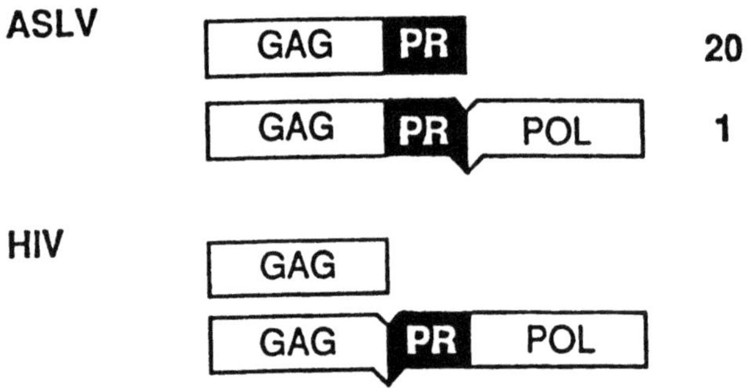

Figure 1 Retroviral Protease (PR) Synthesized as Part of Polyprotein Precursors
In the case of ASLV both gag and gag-pol contain PR whereas in HIV-1 only gag-pol contains PR.

The expression of Pr76*gag*, which contains the PR, is 20 times higher than that of Pr180*gag-pol*. As a consequence the molar ratio of enzyme to target sites in Pr76*gag* is of the order of 1:6 and there is an excess of enzyme over target sites in Pr180*gag-pol* (Fig. 2). In mammalian viruses, where the protease is expressed as part of the gag-pol precursor (Fig. 1), this ratio would be about one tenth. Thus the processing of viral precursors can be carried out by a protease characterized by a very low turnover number and by a very high specificity for its target sites.

The results of our laboratories indicate that i) small synthetic peptides can be used as model substrates to investigate the specificity of the molecular parameters of retroviral polyprotein cleavage by PR. ii) the estimated minimal turnover number of ASLV PR cleavage of the Tyr-Pro decapeptide is 1-3 per min. iii) the substitution of amino acids at either side of the bond to be cleaved affects the ability of a peptide (and larger precursor) to serve as substrate. The specificity seems to be more stringent at the -1 site (Fig. 3)(cf. ref. 14). Cleavage sites in a viral polypeptide can be deduced from a comparison of the amino acids at the C- and N-termini of the mature gag and pol peptides, aligned with the nucleotide sequence of the precursors (Fig. 2) (15-17). Three questions are under study in our laboratory concerning the cleavage sites, namely, i) Are all cleavages in the gag and gag-pol precursors carried out by PR alone or do cellular enzymes participate in this process? ii) Is there a unique order of cleavage along the precursor molecules? and iii) Is there a consensus cleavage sequence for retroviral proteases?

Our preliminary results showed that the ASLV precursors such as NC-PR (18), or NC-PR-RT-IN are stable in E. coli as long as PR is inactive (mutated at the active site). Similar results were obtained previously in eukaryotes by Katoh et al. (1). It seems therefore that the viral precursors are

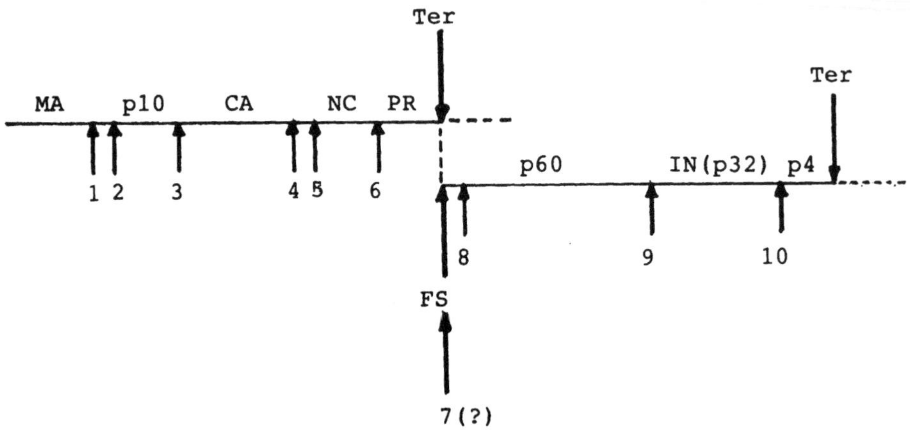

1.	Gly	Thr	Ser	Cys	Tyr--His	Cys	Gly	Thr	Ala
2.	Pro	Pro	Tyr	Val	Gly--Ser	Gly	Leu	Tyr	Pro
3.	Pro	Val	Val	Ala	Met--Pro	Val	Val	Ile	Lys
4.	Ile	Ala	Ala	Ala	Met--Ser	Ser	Ala	Ile	Gln
5.	Gln	Pro	Leu	Ile	Met--Ala	Val	Val	Asn	Arg
6.	Pro	Pro	Ala	Val	Ser--Leu	Ala	Met	Thr	Met
7.	Arg	Leu	Thr	Asn	Leu--Ile	Gly	Arg	Ala	Thr
8.	Arg	Ala	Thr	Val	Leu--Thr	Val	Val	Leu	His
9.	Thr	Phe	Gln	Ala	Tyr--Pro	Leu	Arg	Glu	Ala
10.	Ser	Pro	Leu	Phe	Ala--Gly	Ile	Ser	Asp	Trp

Figure 2 A: Location of Cleavage Sites Believed to Be Hydrolyzed by ASLV PR
FS - frameshift site, designated by dotted line.
TER - termination codon of _gag_ and _gag-pol_ polypeptides. The drawing does not show the actual scale.

B: Amino Acid Sequences Spanning Peptide Bonds Shown in Panel A
The cleavage sites are designated by (--).
Information for this figure is from the work of Schwartz et al. (15), and from our unpublished data.

Peptide No	Sequences										Relative Cleavage Rate
	-5	-4	-3	-2	-1	+1	+2	+3	+4	+5	
1.	Thr	Phe	Gln	Ala	Tyr--	Pro	Leu	Arg	Glu	Ala	1
2.	Thr	Phe	Gln	Ala	Ala--	Pro	Leu	Arg	Glu	Ala	0 (<0.02)
3.	Thr	Phe	Gln	Ala	Phe--	Pro	Leu	Arg	Glu	Ala	0.76
4.	Thr	Phe	Gln	Ala	Ile--	Pro	Leu	Arg	Glu	Ala	0
5.	Thr	Phe	Gln	Ala	Tyr--	Gly	Leu	Arg	Glu	Ala	0.08
6.	Thr	Phe	Gln	Ala	Tyr--	Asp	Leu	Arg	Glu	Ala	0.13
7.		Ac	Gln	Ala	Tyr--	Pro	Leu	Arg	Glu	Ala	0

Figure 3 Peptides Tested as Substrates for ASLV PR and Their Relative Cleavage Rates

quite stable in homologous and heterologous cells suggesting that the cleavage is carried out by PR. It is as yet not known whether the cleavage at one site in the precursor polypeptide depends on the cleavage of other target sites. Introduction of mutations at a specific cleavage site followed by the release of other mature proteins, should solve this problem.

A consensus cleavage sequence for retroviral protease has been proposed on the basis of a comparison of the amino acids adjacent to known sites of human T-cell leukemia virus 1 and MuLV precursor proteins (19). Recently Pearl and Taylor (20) have noted a strongly conserved pattern encompassing amino acids positioned at -3 to +3 with respect to the peptide bond to be cleaved. Our own computer analyses (using methods described by Kyte and Doolittle, Chou and Fasman, and Hopp and Woods) of the sequences of 27 protease cleavage sites from various retroviruses, including all of the known sites in ASLV and HIV, also show a generally hydrophobic pattern. However, no common secondary structure of these sites is evident within or among proteins from different viruses. Certain dipeptide sequences do appear frequently at the cleavage site. For example, Tyr-Pro is found in gag and gag-pol cleavage sites in avian, human, and murine viruses. In fact, dipeptides Tyr-Pro and Phe-Pro comprise approximately 25% of the total cleavage sites examined. Recently Henderson et al. (16) divided the amino acid sequences around peptide bonds believed to be cleaved by HIV-1 and simian immunodeficiency virus PR's into three classes according to similarities in sequences found at both sides of the cleavage site.

However, the sequence variability of target sequences in each of the cleavage sites in the viral precursor proteins suggests that secondary and tertiary structure features are also important for the protease-substrate recognition. This conclusion is supported by the following two findings: i) the HIV-1 protease does not cleave denatured 27 kD gag precursor substrates (21); hence, although this gag fragment contains cleavage sites, a particular conformation may be essential for their recognition, ii) the denatured viral ASLV Pr76^{gag} protein can be cleaved by PR *in vitro*, although some sites do not appear to be efficiently recognized (6).

It is worthwhile to mention that since it is not yet clear whether the processing of the gag-pol polypeptides is carried out solely by the viral protease or whether some cellular proteases take part in this process, the use of synthetic peptides and of purified viral proteases may clarify this question and may also indicate the relative efficiency of cleavage of each peptide by PR.

The Retroviral-Protease is Similar to the Aspartic Proteases

The retroviral protease and the analogous domains in yeast Ty elements, Drosophila Copia-like elements, Cauliflower Mosaic virus and Hepatitis B virus contain highly conserved amino acid sequences (Asp-Thr-Gly) that are identical with the active site sequence of the family of cellular aspartic proteases (also called acid or carboxyl proteases) (22-26). Recently we (18) and others (27-29) have demonstrated that alteration of the Asp residue in this putative active site of the enzyme inactivates the HIV-1 and ASLV PR's. Although there are similarities in the active sites (the active site of ASLV PR contains Asp-Ser-Gly) of viral and cellular proteases, several differences exist between them. The cellular proteases contain more than 300 residues, while the retroviral proteases do not exceed 130 amino acids in length. The molecules of cellular aspartic proteases have two domains and each domain includes the conserved sequence Asp-Thr-Gly (e.g. in pepsin these active aspartates are residues at positions 32 and 215). A model in which the viral enzyme acts as a dimer was introduced by Pearl and Taylor (20). The predicted tertiary structure of the viral enzyme dimer would form an active site with a putative two-fold axis of symmetry, similar to that formed in the cellular aspartic proteases and would thus develop the substrate "cleft" for cleavage sites in retroviruses (23).

Recently we compared the ASLV PR with three cellular proteases, namely with pepsin, renin and cathepsin D and have found (30) that i) the ASLV protease is eluted from a gel filtration column as a dimer, ii) the PR is most active at pH 5.5, at 35°C and in 2-3M NaCl. (At these high salt concentrations the PR looses its specificity and cleaves decapeptides which are resistant at a low ionic strength of the solution. This high salt concentration also increased the proteolytic activity of pepsin), iii) the specificity of ASLV PR for its cleavage sites is more stringent than that of cellular enzymes for the same substrate, and iv) pepstatin, a strong inhibitor of pepsin, renin and cathepsin D had little effect on the PR.

Release of PR from Precursor

The ASLV PR is synthesized as part of the *gag* and *gag-pol* precursors and is released from the polyproteins by cleavage at the N-terminus between NC and PR (15). This cleavage releases the PR from the *gag* precursor, however, additional cleavage should release the PR from precursor Pr180$^{gag-pol}$. The amino acid at the C-terminus of PR released from Pr180$^{gag-pol}$ and the cleavage site in this region have not been defined as yet.

To date it is not clear how PR is cleaved off from the precursor peptides; two plausible alternatives, however, can be proposed: i) Cellular proteases are responsible for the initiation of the cleavage process. Few PR molecules are released and their action releases more PR as required for the final processing of viral polypeptides. ii) PR is released as a result of autocleavage, via an inter- or intra-molecular action.

As yet there is no evidence that cellular protease(s) participate in the release of PR. Evidence showing that viral precursors are processed in heterologous prokaryotic and eukaryotic cells suggests that there is no need for a specific cellular enzyme to initiate the process. On the other hand, we have shown that a) PR is able to control correctly its own cleavage from the neighboring *gag* protein nucleocapsid (NC) expressed as a fused protein in bacteria, and b) NC-PR cleavage is blocked by a mutation in the putative active site of PR. This clearly demonstrates that PR is involved in releasing itself from the *gag* precursors; this does not exclude, however, the possibility that the initiation of the process is effected by a cellular enzyme.

Activation of PR

The release of PR from the precursor proteins may be required for its activation. This mechanism is well known in other aspartic and non-aspartic proteases where an inactive zymogen is converted into the catalytically active enzyme by covalent activation. One example is pepsin (an aspartic protease), which is synthesized as inactive zymogen, pepsinogen. When pepsinogen is exposed to low pH in the gastrointestinal tract it is converted into its active form, pepsin, by hydrolytic cleavage of several specific peptide bonds. In this case the proteolytic activation is carried out by the end product, pepsin (31). On the other hand, trypsinogen and chymotrypsinogen (non-aspartic proteases) are activated by heterologous enzymes, enterokinase and trypsin, respectively. Alternatively, if PR is active in the precursor form, auto- or intercatalytic reactions could result in the release of only PR or of all the mature gag and pol peptides.

Sites of Viral Protease Activity

The activation site of C-type viruses has been extensively studied and the present state of our knowledge can be summarized as follows: i) The core structure and pol polyproteins accumulate close to the membrane region where viral envelope glycoproteins are embedded (3). ii) The gag and pol precursor polypeptides are cleaved into the mature subunits concomitantly with the release of virions into the culture media (1, 3, 8 , 32). iii) Free mature viral proteins are not abundant in infected cells. iv) An active protease can be purified from mature virions (6, 33, 34). v) Viruses mutated in the protease region release particles that have deformed shapes and are not infectious (1, 2). The deficiency of these mutants can be complemented by wild-type viruses (1). These observations strongly suggest that the PR precursor produced in vertebrate host cells, is devoid of proteolytic activity until it reaches the membrane milieu. This delay is most likely essential for the coexistence of the virus and the host cell because it may prevent PR from digesting cellular proteins, block the effect of transcriptase (RT) and endonuclease (EN) activities, which could result in unregulated gene amplification, and because it could provide for migration of viral structural components to the site of virion formation in a suitable molar ratio. Five possible mechanisms could account for PR activation during virion formation: i) The activation of PR occurs under specific conditions present in the cell membrane which do not exist in the cytoplasm, such as a hydrophobic or lipidic environment. ii) The cellular protease(s) present in the cell membrane may be responsible for the release of PR or alternatively for modifying the precursor thus enabling the initiation of processing by PR. iii) The activation of PR requires high local concentrations for intermolecular reactions or dimerization to occur (20). iv) The protein-protein interaction at the precursor level during assembly of the mature virion initiates PR activation. v) Meristylation, phosphorylation or other posttranslational modifications of precursor molecules may inhibit the processing of the poly-proteins in the cell cytoplasm. Attachment to the cell membrane may somehow relieve this inhibition.

The finding that retrovirus precursor proteins undergo processing in a variety of heterologous cells namely ASLV in bacteria (18, 35) and Xenopus laevis oocytes (36), HIV-1 in yeast (12) and in bacteria (21, 37, 38) are inconsistent with mechanisms requiring specific proteases or specific conditions in the cell membrane.

Clarification of the regulation mechanisms controlling the activation of PR may be of great importance for the synthesis of suitable inhibitors of retroviral PR.

Acknowledgments

I thank Dr. A.M. Skalka, in whose laboratory a part of this work was done, for her collaboration, her valuable comments and her continuous support. I also thank Dr. Thea Pugatsch for her help in editing this manuscript.

References

1. Katoh, I., Y. Yoshinaka, A. Rein, M. Shibuya, T. Okada S. Oroszlan. 1985. Virology 145, 280-292.

2. Kohl, N.E., E.A. Emini, W.A. Schleif, L.J. Davis, J.C. Heimbach, R.A. Dixon, E.M. Scolnick, I.S. Sigal. 1988. Proc. Natl. Acad. Sci. USA 85, 4686-4690.

3. Weiss, R., N. Teich, H. Varmus, J. Coffin. 1984. In: Molecular Biology of Tumor Viruses; RNA tumor viruses, 2nd ed., Cold Spring Harbor, New York.

4. Von der Helm, K. 1977. Proc. Natl. Acad. Sci. USA 74, 911-915.

5. Dittmar, K.J., K. Moelling. 1978. J. Virol. 28, 106-118.

6. Vogt, V.M., W. Wight, R. Eisenman. 1979. Virology 98, 154-167.

7. Eisenman, R.N., W.S. Mason, M. Linial. 1980. J. Virol. 36, 62-78.

8. Voynow, S.L., J.M. Coffin. 1985. J. Virol. 55, 79-85.

9. Witte, O.N., D. Baltimore. 1976. J. Virol. 26, 750-761.

10. Levin, J.G., S.C. Hu, A. Rein, L.I. Messer, B.I. Gerwin. 1984. J. Virol. 51, 470-478.

11. Crawford, S., S.P. Goff. 1985. J. Virol. 53, 899-907.

12. Kramer, R.A., M.D. Schaber, A.M. Skalka, K. Ganguly, F. Wong-Staal, E.P. Reddy. 1986. Science 231, 1580-1584.

13. Jacks, T., H.E. Varmus. 1985. Science 230, 1237-1242.

14. Kotler, M., R.A. Katz, W. Danho, J. Leis, A.M. Skalka. 1988. Proc. Natl. Acad. Sci. USA 85, 4185-4189.

15. Schwartz, D.E., R. Tizard, W. Gilgert. 1983. Cell 32, 853-869.

16. Henderson, L.E., R.E. Benveriste, R. Sowder, T.D. Copeland, A.M. Schultz, S. Oroszlan. 1988. J. Virol. 62, 2587-2595.

17. Lillehoj, E.P., F.H.R. Salazar, R.J. Mervis, M.G. Raum, H.W. Chan, N. Ahmad, S. Venkatesan. 1988. J. Virol. 62, 3053-3058.

18. Kotler, M., R.A. Katz, A.M. Skalka. 1988. J. Virol. 62, 2696-2700.

19. Oroszlan, S., T.D. Copeland. 1985. Curr. Top. Microbiol. Immunol. 115, 221-233.

20. Pearl, L.H., W.R. Taylor. 1987. Nature 328, 482.

21. Hansen, J., S. Billich, T. Schulze, S. Sukrow, K. Moelling. 1988. EMBO J. 7, 1785-1791.

22. Toh, H., M. Ono, K. Saigo, T. Miyata. 1985. Nature 315, 691.

23. Pearl, L.H., W.R. Taylor. 1987. Nature 329, 351-354.

24. Toh, H., R. Kikuno, H. Hayashida, T. Miata, W. Kugimiya, S. Inouye, S. Kuki, K. Saigo. 1985. EMBO J. 4, 1267-1272.

25. Power, M.D., P.A. Marx, M.L. Bryant, M.B. Gardner, P.J. Barr, P.A. Luciw. 1986. Science 231, 1567-1573.

26. Miller, R.H. 1987. Science 236, 722-725.

27. Debouck, C., J.G. Gorniak, J.E. Strickler, T.D. Meek, B.W. Metcalf, M. Rosenberg. 1987. Proc. Natl. Acad. Sci. USA 84, 8903-8904.

28. Le Grice, S.F.J., J. Mills, J. Mous. 1988. EMBO J. 7, 2547-2553.

29. Mous, J., E.P. Heimer, S.F.J. Le Grice. 1988. J. Virol. 62, 1433-1436.

30. Kotler, M., W. Danho, R.A. Katz, J. Leis, A.M. Skalka. 1988. J. Biol. Chem., in press.

31. Tang, J. 1979. Mol. Cell. Biochem. 26, 93-109.

32. Lu, A.H., M.M. Soong, P.K.Y. Wong. 1979. Virology 93, 269-274.

33. Moelling, K., A. Scott, K.E.J. Dittmar, M. Owwada. 1980. J. Virol. 33, 680-688.

34. Alexander, F., J. Leis, D.A. Soltis, R.M. Crowl, W. Danho, M.S. Poonian, Y.-C.E. Pan, A.M. Skalka. 1987. J. Virol. 61, 534-542.

35. Mermer, B., M. Malamy, J.M. Coffin. 1983. Mol. Cell. Biol. 3, 1746-1748.

36. Katz, R.A., G.M. Maniatis, R.V. Guntaka. 1979. Biochem. Biophys. Res. Commun. 87, 447-453.

37. Framerie, W.G., D.D. Leob, N.C. Casavant, C.A. Hutchinson III, H.E. Marshall, R. Swanstrom. 1987. Science 238, 305-308.

38. Graves, M.C., J.J. Lim, E.P. Heimer, R.A. Kramer. 1988. Proc. Natl. Acad. Sci. USA 85, 2449-2453.

CHARACTERIZATION OF p15 PROTEASE OF MYELOBLASTOSIS ASSOCIATED VIRUS BY SPECIFICITY AND INHIBITION STUDIES

P. Štrop, J. Konvalinka, I. Bláha, L. Pavlíčková, M. Souček, J. Velek, I. Pichová, M. Fusek and V. Kostka

Institute of Organic Chemistry and Biochemistry, Czechoslovak Academy of Science, 166 10 Prague

F. Kaprálek, J. Sedláček

Institute of Molecular Genetics, Czechoslovak Academy of Science, 166 37 Prague

Introduction

As with other retroviruses the maturation of MAV, the avian myeloblastosis-associated virus, is a result of processing of viral polyprotein precursors by limited proteolysis effected by the virus encoded protease (1). The development of a procedure for bacterial expression of the protease (2) as well as for its isolation (3), described elsewhere in this volume, have paved the way to obtaining the enzyme in quantities sufficient for a large scale testing of its specificity with well defined peptide substrates. Such studies are expected to provide data necessary for a rational design of inhibitors of the protease based on substrate analogs. Experiments following this line of approach and using protein substrates and custom-synthesized peptides and inhibitors are described in this study.

Proteases of Retroviruses

Results

Specificity and Mechanism of Action of p15<u>gag</u> Protease of MAV

Hydrolysis of high molecular weight substrates

In orienting experiments the cleavage of 15 various polypeptide substrates of M_r=6-70 kD by recombinant p15<u>gag</u> protease (3) was carried out and the digests were examined by SDS PAGE. Solutions of either native or denatured (by heat, by reduction and carboxymethylation or aminoethylation, by treatment with urea or guanidinium chloride) substrates, such as basic pancreatic trypsin inhibitor, soybean trypsin inhibitor, porcine and chicken pepsin and pepsinogen, lysozyme, ovalbumin and human hemopexin, were digested in conventional buffers at pH 5.0 at protease to substrate ratios ranging up to 1:5; no cleavage, however, was detected. These experiments were successful with heat denatured human or bovine serum albumin (HSA or BSA, resp.) only on condition that the assays contained SDS and 2-mercaptoethanol (2-ME), as specified in Fig. 1. The pH-optimum of cleavage of HSA was 5.0 and an identical value was obtained also with BSA. During the cleavage partial precipitation of the protein was observed. This showed that the proteins were cleaved virtually in precipitated or at least in aggregated form. This conclusion was confirmed by an experiment in which the reaction mixture, as specified in Fig. 1, after the addition of p15 was centrifuged, the pellet was washed 3 times with the reaction buffer and both the supernatant and the pellet were incubated 2 h at 40°C. The presence of p15 and the cleavage of the substrate were examined afterwards. All of the p15 protease was detected in the pellet fraction and no cleavage products were found in the supernatant fraction (Fig. 2). Our conclusion on the preference of the p15 protease for aggregated substrates received further support

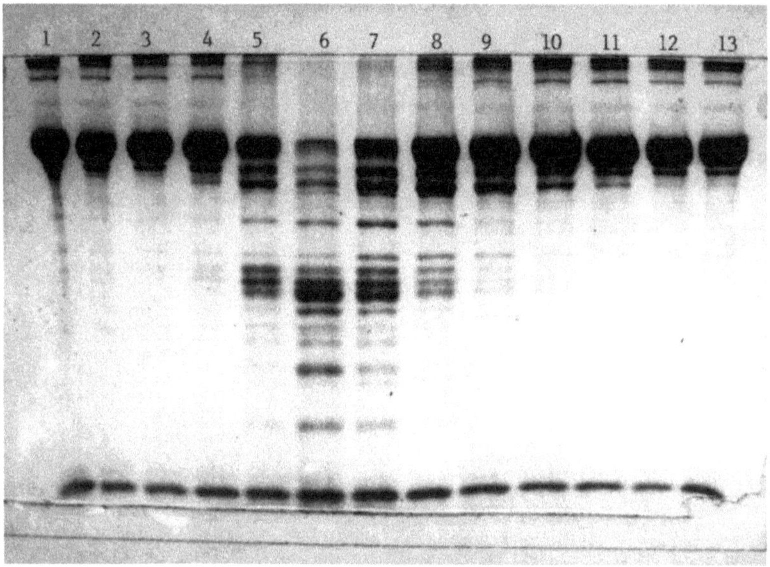

Figure 1 Cleavage of HSA by Recombinant p15^{gag} Protease of MAV
A 0.015 mM solution of the heat denatured protein in buffer at the pH shown was digested by a 0.002 mM solution of the protease 2 h at 40°C in the presence of 0.03 % SDS and 0.05 % 2-ME. Shown are SDS PAGE patterns of digests obtained at pH : 2.0 (lane 1), 3.0 (2), 3.5 (3), 4.0 (4), 4.5 (5), 5.0 (6), 5.5 (7), 6.0 (8), 6.5 (9), 7.0 (10), 7.5 (11), 8.0 (12), and 8.5 (13).

from our digestion experiments with tubulin, a protein from cytoskeleton of M_r=52,000 isolated from pig brain. Unlike the other polypeptide substrates tested tubulin is cleaved undenatured and in the absence of SDS and 2-ME yet the formation of aggregates during the cleavage, which is optimal at pH 6, can be observed. The time profile of the digestion of tubulin is shown in Fig. 3. It remains to be shown whether the preference of the p15 protease for aggregated substrates

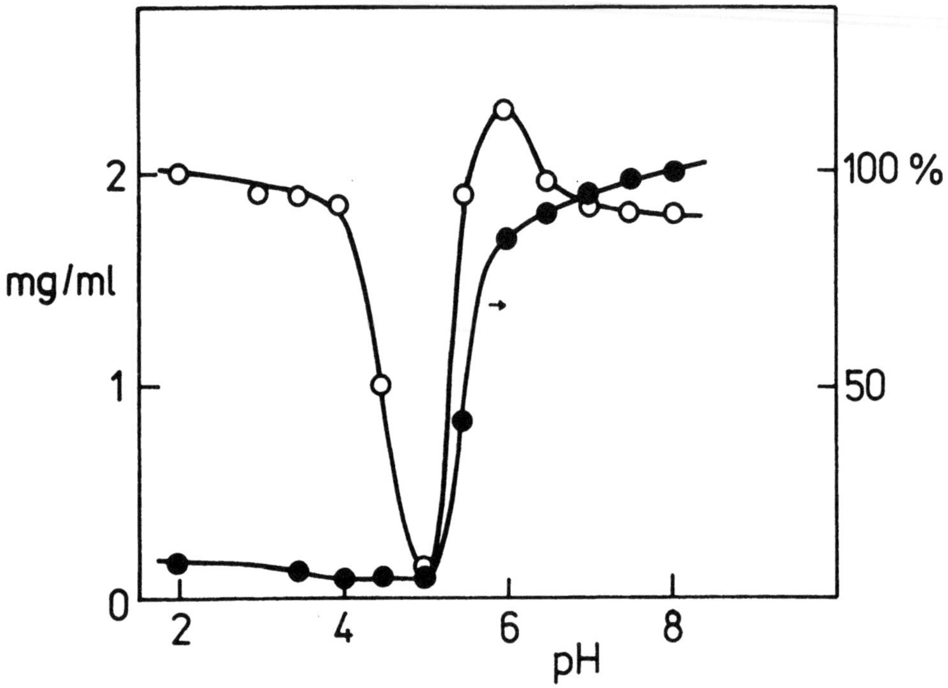

Figure 2 pH-Profile of Cleavage of HSA by p15<u>gag</u> protease
Full symbols-quantity of HSA in solution, open symbols-degree of hydrolysis of HSA. The conditions of the assay are described in Fig. 1.

observed in our *in vitro* studies somehow reflects its action *in vivo* where viral proteases have been hypothesized to act in immobilized form, *i. e.* remaining bound to the cell membrane (4).

Hydrolysis of Synthetic Peptide Substrates

As a first step toward the determination of the specificity of the recombinant p15*gag* protease the frequencies of

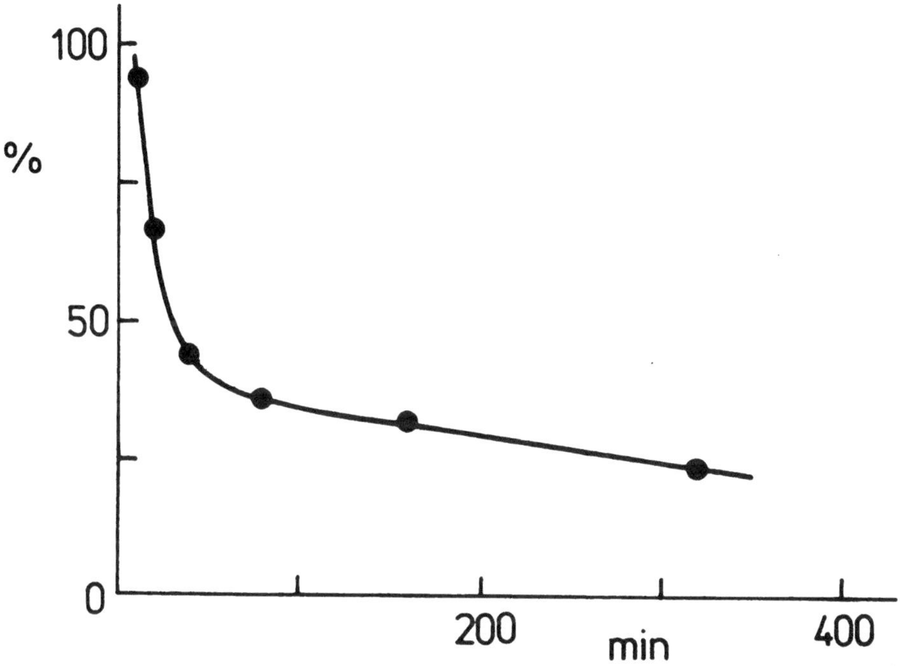

Figure 3 Time Profile of Digestion of Pig Tubulin by p15<u>gag</u> Protease
A 0.022 mM solution of the protein was digested by a 0.003 mM solution of the p15 protease at pH 6.0 and 40°C. Abscissa - % of tubulin digested.

individual amino acid residues in sequences spanning bonds cleaved by the highly homologous Rous sarcoma virus (RSV) protease (5) during the processing of RSV polyproteins were examined. The recorded data were complemented by information on bonds cleaved in HSA and BSA, as determined in our experiments after electrophoretic separation of the cleavage products and their sequencing. The results of the examination are shown in Fig. 4. From the data some homologies in sequences spanning bonds cleaved in the virus polyproteins and in HSA or BSA can be spotted. Homologous are, *e. g.*, the

subsite																bond cleaved	
P								P'								boundary	sequence
8	7	6	5	4	3	2	1	1	2	3	4	5	6	7	8		
K	T	**V**	G	T	S	C	Y	H	C	G	**T**	A	I	G	C	p19/p2a	155-156 gag
T	A	I	G	C	N	C	**A**	T	A	**S**	A	**P**	P	**P**	**P**	p2a/p2b	166-167 gag
A	S	**A**	**P**	**P**	P	P	Y	V	G	**S**	G	**L**	Y	**P**	S	p2b/p2c	175-176 gag
A	**P**	**P**	P	P	Y	V	G	S	G	L	Y	**P**	S	**L**	A	p2c/p10	177-178 gag
T	G	P	**P**	V	**V**	A	**M**	P	**V**	V	I	K	T	E	G	p10/p27	239-240 gag
D	Q	G	I	A	A	A	**M**	**S**	S	A	I	Q	P	**L**	V	p27/p3	479-480 gag
T	M	R	E	K	**V**	L	T	**S**	S	A	R	Q	R	**L**	R		189-190 BSA*
S	A	I	Q	**P**	L	**V**	**M**	A	**V**	V	N	R	**E**	R	D	p3/p12	488-489 gag
G	P	E	**P**	**P**	A	**V**	S	L	A	M	**T**	M	**E**	H	K	p12/p15	577-578 gag
R	L	**V**	R	**P**	E	**V**	D	V	M	C	**T**	A	F	H	D		121-122 HSA*
S	Q	**A**	T	F	Q	A	Y	P	L	R	E	A	K	D	L	p63/p32	572-573 pol
D	**E**	**A**	S	**P**	L	F	**A**	G	I	**S**	D	W	I	**P**	W	p32/p4	858-859 pol
C	**E**	**V**	T	G	**V**	R	**A**	D	**V**	H	L	**L**	E	Q	**P**	p7/p40	64-65 env
L	**E**	K	S	H	C	I	**A**	E	**V**	E	K	D	A	I	**P**		289-290 BSA*
R	T	G	I	R	R	K	R	**S**	**V**	**S**	H	**L**	D	D	T	p40/p23	405-406 env
C	F	S	A	L	E	**V**	D	E	T	Y	V	**P**	K	E	F		494-495 HSA*
T	**E**	**V**	**P**	**P**	**V**	**V**	**A**	**S**	**V**	**S**	**T**	**L**	**E**	**P**	**P**	"IDEAL SUBSTRATE"	

Figure 4 Bonds Cleaved by p15 Protease of RSV or MAV in Virus Polyproteins and HSA or BSA
Data obtained in this study are marked by an asterisk. Preferred residues in the individual subsites are set in boldface.

sequences around the p12/p15 boundary of the MAVgag polyprotein and the sequence around the bond between residues 121 and 122 of HSA (subsites P4, P2, and P4'). While no final distinct conclusions on the preference of p15 for certain amino acids in positions around the cleaved bond can be made as yet the data on the frequencies of the individual residues in these positions were used to advantage for the design of an "ideal substrate" also shown in Fig. 4. Using the Merrifield resin (6) in the classical batchwise procedure two such substrates were synthesized in which the P1 residue was replaced by a p(NO_2)Phe residue so that the cleavage of the scissile bond might be assayed spectrophotometrically. Together with these two substrates a series of 16 octadecapeptides were also synthesized which corresponded to sequences flanking scissile bonds in MAV and RSV polyproteins and in HSA and BSA. These peptides were hydrolyzed by the p15$^{\underline{gag}}$ protease under the conditions shown in Fig. 5 and the digests were resolved by HPLC on a C-18 reverse phase column using the system MeOH/H_2O containing 0.05% TFA, by TLC on silica gel in four solvent mixtures, namely sec. butanol : formic acid : water (75:12.5:12.5), sec. butanol : ammonium hydroxide : water (85:7.5:7.5), n-butanol : pyridine : water (9:1:1) and butanol : pyridine : acetic acid : water (15:10:6:3), and by high voltage paper electrophoresis (7). The products were identified by N-terminal end-group and sequence analysis. As shown in Fig. 5, the highest degree (1.and 2.) of cleavage (almost to 100%) was observed with both "ideal substrates", according to our expectation. The degree of cleavage of 8 additional peptides varied between 90 and 10% (3.-8.). The remaining 8 peptides, representing 6 cleavage sites in virus polyproteins and 1 site in both BSA and HSA (198-199 and 121-122, resp.) were not cleaved at all. The fact that these sites were not accepted as substrates for cleavage is surprising and may reflect possible secondary structure preferences of the protease. With two peptides the main cleavage was between residues in P2' and P3' whereas the putative scissile bond was cleaved to a much lesser degree. Seven substrates of pepsin and chymosin, two general aspartic proteases, were not cleaved at all.

subsite P									P'									degree of cleavage	representing sequence
9	8	7	6	5	4	3	2	1	1'	2'	3'	4'	5'	6'	7'	8'	9'		
	T	T	V	G	P	V	V	**F**│	S	V	S	T	K	E	L	P		2.	ideal substrate
				G	P	V	V	**F**│	S	V	S	T	K	E	L	P		1.	ideal substrate
	K	T	V	G	T	S	C	Y	H	C	G	T	A	I	G	C			P19/P2a
	T	A	I	G	C	N	C	A	T	A	S	A	P	P	P	P			P2a/P2b
	A	S	A	P	P	P	P	Y¦	V	G│	S	G	L	Y	P	S		5.	P2b/P2c
	A	C	A	P	P	P	P	F¦	V	G│	S	G	L	Y	P	S		5.	(P2b/P2c)
	T	G	P	P	V	V	I	M│	P	V	V	I	K	T	E	G		6.	P10/P27
	D	Q	G	I	A	A	A	M	S	S	A	I	Q	P	L	I			P27/P3
	T	M	R	E	K	V	L	T	S	S	A	R	Q	R	L	R			BSA 198-190
	S	A	I	Q	P	L	I	M	A	V	V	N	R	E	R	D			P3/P12
	G	P	E	P	P	A	V	S│	L	A	M	T	M	E	H	K		7.	P12/P15
	G	P	E	P	P	A	V	S│	L	A	M	T	M	D	H	K		7.	(P12/P15)
	S	Q	A	T	F	Q	A	Y│	P	L	R	E	A	V	I	L	K	3.	(P63/P32)
K	D	E	A	S	P	L	F	A	G	I	S	D	T	I	P				P32/P4
	C	E	V	T	G	V	R	A│	D	V	H	L	L	E	Q	P		8.	gp85s
	L	E	K	S	H	C	I	A│	E	V	E	K	D	A	I	P		4.	BSA 289-290
	R	T	G	I	R	R	K	R	S	V	S	H	L	D	D	T			gp37
	R	L	V	R	P	E	V	D	V	M	C	T	A	F	H	D			HSA 121-122
		H	P	H	P	L	S	F	M	A	I	P	P	K	K				chymosin substr.
				K	P	A	E	F	**F**	A	L								pepsin substr.
				K	P	L	E	F	**F**	R	L								chymosin substr.
				G	L	L	E	D	**F**	L	K	K	Q						chymosin substr.
					K	Y	H	P	**F**	L	T	A							pepsin substr.
				K	P	A	A	F	**F**	A	A	Amide							pepsin substr.
	F	G	H	P	H	L	S	F	M	A	I	P	P	K	R(DNS)				chymosin substr.

Figure 5 Cleavage of Synthetic Peptide Substrates by p15*gag* Protease of MAV
Solutions (0.1mM) of the substrates in 0.2M phosphate and 0.1M citrate buffer, pH 6.0, containing 0.001M EDTA, were digested with the protease at enzyme to substrate ratios ranging between 1:40 and 1:80 at 38°C for 5 h. Primary and secondary cleavage is marked by a full and dashed vertical bar, resp. **F** = p(NO_2)Phe.

Table 1 Protease Inhibitors Tested with p15*gag* Protease of MAV

Inhibitor	Concentration, M
Pepstatin A	2.5×10^{-5}
Peptide VLPFFVL*	4.0×10^{-4}
Peptide VFPFFVL*	4.0×10^{-4}
Peptide H142**	2.0×10^{-4}
Leupeptin	2.0×10^{-4}
Cystatin	1.4×10^{-5}
Basic pancreatic trypsin inhibitor	3.0×10^{-4}
Soybean trypsin inhibitor	5.0×10^{-5}
p-aminobenzamidine	5.0×10^{-4}
EDTA	1.0×10^{-3}
Cu^{2+}	1.0×10^{-4}

* The D-amino acids in the substrate-based inhibitors (*cf.* 8) of aspartic proteases are underlined. ** A synthetic renin inhibitor (*cf.* 9).

Inhibition Studies on p15*gag* Protease of MAV

Effect of various protease inhibitors

In orienting experiments the effect of inhibitors of general proteases of the 4 main classes was examined (Table 1). The preincubation of the protease with any of these inhibitors at the concentration shown did not inhibit its activity as assayed with denatured HSA as substrate under the conditions described in Fig. 1. By contrast after 19-70 h of preincubation of a 2/um solution of the p15 protease with 12 mM 1,2 epoxy-3-(p-nitrophenoxy)propane (EPNP), a specific pepsin inhibitor (10), no cleavage of HSA was observed in the standard assay (11).

Effect of pepstatin and substrate analogs

The absence of inhibition of the $p15^{gag}$ protease by pepstatin A in the standard assay as well as the failure of the protease to bind to a pepstatin-Sepharose column under the conditions described for the bovine leukemia virus protease by the Japanese authors (12) led us to test the effect of various pepstatin and reduced bond substrate analogs which are shown in Table 2. These analogs were based on the results of specificity testing of synthetic substrates. Of the latter, five which were cleaved best by the protease (Fig. 5) were chosen for modification. In these substrates two residues of the putative scissile bond were replaced by statine, statine derivatives or reduced bond analogs shown in Fig. 6.

Table 2 Inhibition of $p15^{gag}$ Protease of MAV by Pepstatin and Substrate Analogs

No	inhibitor	type	inh. degree
1	Gly-Pro-Val-Val-**PheSta**-Val-Ser-Thr-Ala	statine	0
2	Pro-Pro-Pro-Pro(PheVal)Gly-Ser-Gly-Leu	reduced bond	0
3	Pro-Pro-Pro-Pro-**PheSta**-Gly-Ser-Gly-Leu	statine	0
4	Pro-Pro-Ala-Val-**CysSta**-Ala-Met-Thr-Met	statine	70%
5	Thr-Phe-Gln-Ala-**PheSta**-Leu-Arg-Glu-Ala	statine	90%
6	Thr-Phe-Glu-Ala(TyrPro)Leu-Arg-Glu-Ala	reduced bond	90%
7	Pro-Pro-Val-Val---**Sta**--Pro-Val-Val-Ile	statine	0

The inhibition was tested with peptides GPVVFSVSTKELP and SQATFQAVPLREAVILK as substrates using the assay system described in Fig. 5. A 1 mM solution of the protease was incubated with the inhibitors and substrates for 6 h. Under such conditions a 70 or 90% inhibition was observed with inhibitors 4 and 5 and 6, resp.

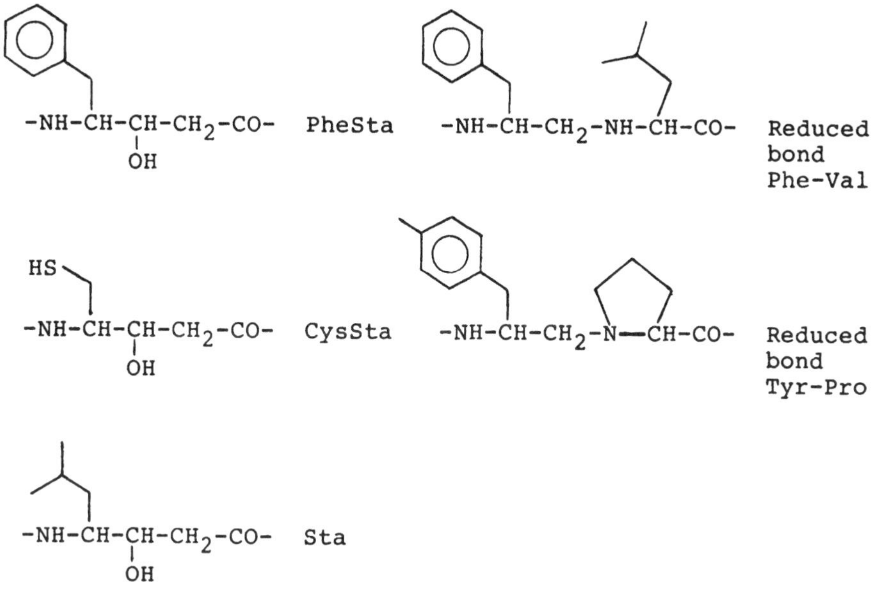

Figure 6 Statine Derivatives and Reduced Bond Substrate Analogs

Inhibition was observed with analogs 4, 5 and 6 only. Inhibitors of general aspartic proteases (cf. Table 1) or other statine-based inhibitors, synthesized in our Laboratory, such as Lys-Pro-Leu-Glu-**PheSta**-Arg-Leu-Pro, a chymosin inhibitor, or inhibitors successfully used with the HIV-1 protease (13), were without any effect on the recombinant p15^{gag} protease of MAV.

Identity of p15^{gag} Protease of MAV

It was postulated in one of the pioneer studies (11) on the biochemical characteristics of the "proteolytic activity associated with p15^{gag} of MAV" that this activity has the features of a cysteine protease. This hypothesis can hardly withstand the less ambiguous experimental evidence available today. The postulate that this protease is an aspartic protease, originally based only on the occurrence in its primary structure of tripeptide sequence Asp-Thr-Gly identical to the active site sequence of aspartic proteases, receives further support from the results of inhibition studies with EPNP and pepstatin analogs reported here. The protease appears to recognize similar sequence patterns as other aspartic proteases. We have not been able to show (14), however, that the p15 protease forms dimers without which it could not act as a competent aspartic protease as follows from the results of molecular modeling. The hypothesis that the p15 protease of MAV is an aspartic protease received additional experimental proof from crystallographic studies during the preparation of this manuscript (15).

Acknowledgment

The skillful technical assistance of Dr. M. Pavlík, and Mr. Zbrožek in the experimental part of this study is gratefully acknowledged.

References

1. Vogt, V.M., R. Eisenman, H. Diggelmann. 1975. J. Mol. Biol. 96 471-493.

2. Sedláček, J., F. Kaprálek, P. Štrop, I. Pichová, V. Kostka, M. Trávníček. 1989. This volume, pp. 49-59 .

3. Štrop, P., I. Pichová, V. Kostka, F. Kaprálek, J. Sedláček. 1989. This volume, pp. 61-72.

4. Haseltine W.A., F. Wong-Staal. 1988. Sci. Am., Oct., 52-62.

5. Schwartz, D.E., R. Tizard, V. Gilbert. 1983. Cell 32, 853-869.

6. Merrifield, R.B. and D.W. Wooley. J. Am. Chem. Soc. 78, 4646-4649.

7. Mikeš, O. 1957. Collect. Czech. Commun. 32, 831-850.

8. Pohl, J., M. Baudyš, V. Kostka. 1983. Anal. Biochem. 133, 104-110.

9. Szelke, M. 1985. in Aspartic Proteinases and Their Inhibitors (V. Kostka, ed.), pp. 421-441, W. de Gruyter, New York-Berlin.

10. Chen K.C.S., J. Tang. 1972. J. Biol. Chem. 247, 2566-2574.

11. Dittmar, K.J., K. Moelling. 1978. J. Virol. 28, 106-118.

12. Katoh, I., T. Yasunaga, Y. Ikawa, Y. Yoshinaka. 1987. Nature 329, 654-656.

13. Moelling, K., M.T. Knoop, S. Billich, I. Blaha, L. Pavlíčková, M. Souček. 1989. This volume, pp. 155-164.

14. Kostka, V., P. Štrop, J. Konvalinka, I. Pichová, M. Fusek, I. Bláha, J. Velek, L. Pavlíčková, M. Souček, J. Urban, J. Sedláček, M. Fábry, F. Kaprálek, M. Trávníček in Proceedings of 7th International Symposium on Intracellular Protein Catabolism (Katunuma, N. and Kominami, E., eds.), Springer Verlag and Japan Sci. Soc. Press, Berlin-Tokyo, in press.

15. Miller, M., M. Jaskolski, M.R.J.K. Rao, J. Leis, A. Wlodawer. 1989. (personal communication)

CHARACTERIZATION OF HIV-1 PROTEASE SUBSTRATE SPECIFICITY BY USE OF SYNTHETIC PEPTIDES

K. Moelling, M.T. Knoop, S. Billich,
Max-Planck-Institut für Molekulare Genetik, Abt. Schuster, Ihnestrasse 73, D-1000 Berlin 33, Germany

I. Bláha, L. Pavlíčková, M. Souček
Institute of Organic Chemistry and Biochemistry, Czechoslovak Academy of Science, Flemingovo 2, 166 10 Prague, Czechoslovakia

Introduction

The human immunodeficiency virus HIV-1 is the causative agent of the acquired immunodeficiency syndrome AIDS and is a member of the retrovirus group (1,2). Molecular organization of the HIV-1 genome resembles that of other retroviruses and comprises gag, pol and env as genes necessary for viral replication (3). Translation of the viral genes yields precursor polypeptides that are cleaved by a viral protease to produce the mature forms of the proteins found in virions (4). Processing of gag- and pol- but not of env precursors is assumed to be mediated by the retroviral protease (5). Several groups have recently described the bacterial expression of HIV-1 protease which arises by autocatalytic processing of larger polyprotein precursors (6-10). The protease is an aspartic protease of 9 kDa as was evidenced by use of a monoclonal antibody and demonstration of its inhibition by pepstatin A (6). The protease forms dimers (6). The carboxy-terminal 17 amino acids are essential for the enzyme activity because their deletion results in enzyme inactivation (6). Here we use short synthetic peptides as substrates and inhibitors for

Proteases of Retroviruses

the partially purified bacterially expressed HIV-1 protease to characterize its sequence specificity. The peptides have been selected from the putative cleavage sites of the gag- and pol- precursors.

Results

Synthetic Peptides as Substrates and Inhibitors

For substrate analysis a partially purified bacterially expressed HIV-1 protease was used. A BglII/EcoRV construct was expressed using a trp9 expression vector and the HB101 E.coli strain. The purification procedure comprised a polyethylenimine precipitation, a fractionated ammonium sulfate precipitation, and a gel filtration step (Ultrogel 54, Pharmacia). Details have been described (6). The purification procedure was monitored by a monoclonal antibody (cl 5/18) prepared against the HIV-1 protease using immunoblotting technique (6). Purification of bacteria not expressing the HIV-1 protease did not exhibit any protease activity. The protein content as judged from polyacrylamide gels and silverstain amounts to approximately 1 µg/ml. The partially purified protease actively cleaved gag precursor proteins in vitro (6).

For determination of substrate specificity we selected synthetic model peptides from the putative gag and pol cleavage sites. The lengths of the peptides varied between 7 and 18 amino acid residues. The model peptides were incubated with the partially purified protease for about 1 hr at 37°C using standard assay conditions (6) and 2 to 5 mM concentrations of the peptides. The reaction products were analyzed by reversed phase HPLC using a peptide TS column and eluted with a linear acetonitrile gradient in 50 mM phosphoric acid. The result of a typical analysis of a model peptide before and after cleavage is illustrated in Fig 1.

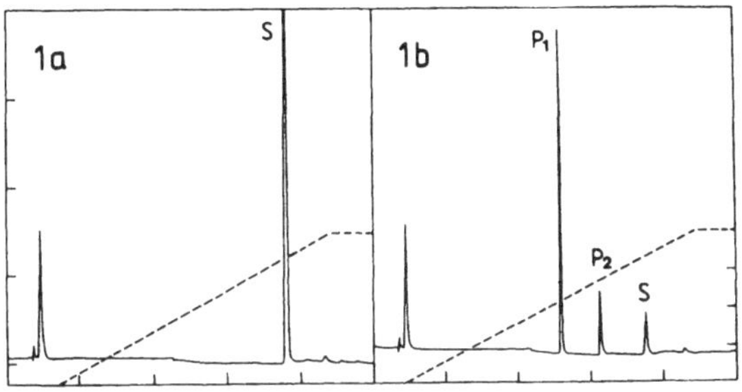

Fig. 1 HPLC profile of synthetic peptide cleaved by HIV-1 protease.
The synthetic peptide SSQVSQNY'PIVQNLGK was cleaved by partially purified HIV-1 protease in vitro, 25 μl assay containing 15 mM MES buffer (morpholinoethane sulfonic acid), pH 6.0, 0.08% (v/v) Triton X-100, 4 mM EDTA, pH 7.0, 0.5 to 2.0 mM peptide substrate, 5 μl of Ultrogel ACA54 fractionated HIV-1 protease, incubated for 2 hrs, 37°C). The cleavage products were analyzed on HPLC column (Peptide TS, 240 x 4.6 mm) and eluted with a linear gradient of acetonitrile in 50 mM phosphoric acid. The cleavage products (P_1) and (P_2) and the uncleaved substrate (S) are indicated. 1a and b indicate the HPLC profiles of the substrate before and after cleavage.

Five gag and pol putative cleavage sites were analyzed using peptides of two sizes, 17 to 18 and 7 amino acid residues in length. Both sizes were cleaved efficiently. The peptides used are listed in Table 1. The position of the cleavage sites on the gag and pol precursors is also indicated in Table 1. Some other heptapeptides were analyzed as well. One of them represented a hypothetical cleavage site between reverse transcriptase and RNase H. This site was not cleaved. Also, only one of two putative cleavage sites between the RNase H and endonuclease was accepted as substrate for cleavage. Three of the peptides contained the

peptide	cleavage	position
SSQVSQNY PIVQNLGK··	yes	gag p17/p24
SQNY PIV	+++	
·GHKARVL AEAMSQVTNP	yes	gag p24/p7
ARVL AEA	+++	
PGNF LQS	++	gag p7/p6
QGTVSFNF PQITLWQK··	yes	prot, N-term
SFNF PQIT	+++	
SFNF PQI	+++	
QIGCTLNF PISPIETK··	yes	prot/RT
TLNF PIS	++	prot/RT
LEKE PIV	—	RT/RNase H
··GIRKIL FLDGIDK···	yes	RNase H/endo
RKIL FLD	+	RNase H/endo
DFNL PPVV	—	RNase H/endo

inhibitor	50% inhibition
VSFN(phe-pro)QITL	< 125 μM
CTLN(phe-pro)ISPI	125 μM
TATI - sta - QRGN	< 125 μM
VSFN(phe-sta)QITL	< 125 μM

Table 1 List of substrates and inhibitors.
The table shows the synthetic peptides analyzed. Cleavage is indicated by yes (when quantitation of cleavage product was complicated by overlapping intermediates from peptide synthesis) or by + and - , with +++ representing 50 to 85%, ++ 20 to 30%, and + 10 to 20% of cleavage. The numbers were derived from integration of the areas of the peaks P_1 and P_2. Abbreviations: RT: reverse transcriptase, N-term: amino terminus, prot: protease, endo: endonuclease.

dipeptide sequence tyr-pro (Y-P) or phe-pro (F-P) as predicted cleavage site (11). Amino acid analysis of the cleavage products of SQNYPIV, SFNFPQI and TLNFPIS confirmed that cleavage takes place next to the proline (data not shown). However, it is known that other peptides are cleaved as well, e.g. methionine-methionine, indicating that proline is not essential.

Four peptides which were cleaved by the protease were modified. In two cases the bond at the cleavage site was

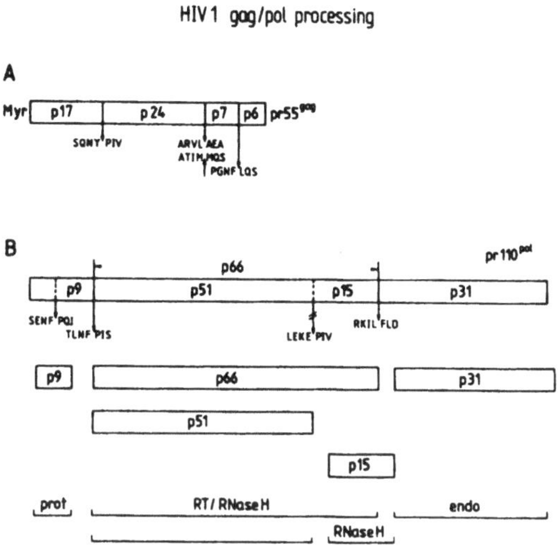

<u>Fig. 2</u> Polyprotein precursors of HIV-1 with indicated cleavage sites.

The gag precursor $p55^{gag}$ and the pol precursor $pr110^{pol}$ are indicated. $Pr110^{pol}$ arises from a gag-pol precursor (not shown) which is cleaved at the amino terminus of the protease P9. $Pr55^{gag}$ is myristilated (Myr), the carboxy-terminal amino acids p7 and p6 arise from a p15 precursor (13). Two cleavage sites between p24 and p7 occur (13). The protease arises by autocatalytic processing at the indicated cleavage sites as predicted (11). No cleavage occurs between the reverse transcriptase molecules p66 and the p15 RNase H which therefore may be attributed to a cellular protease. The cleavage product p15 was shown to exhibit RNase H activity as does the intact p66 polyprotein (6).

reduced and in two other cases the amino acid statine was introduced. Statine acts as a non-hydrolyzable dipeptide analogue resistant to attack of proteases (12). All three peptides inhibited cleavage of the model peptide SQNYPIV. 50% inhibition occurred around 125 µM concentrations (Table 1).

The results obtained with synthetic peptide substrates are illustrated in Fig. 2, where the gag and pol precursor proteins are shown schematically and the cleavage sites indicated.

Using the synthetic substrate SQNYPIV we determined some biochemical properties of the protease. Quantitation of the cleavage efficiency was performed by integration of the areas of S, P_1 and P_2. With this approach we determined the optimum incubation temperature 37°C, the optimum pH ranging broadly from 3.5 to 5.5. The kinetics of the protease reached a plateau after 1 hr at 37°C and stayed at that level for at least 18 hrs. An optimum concentration of NaCl was not observed, the activity was not significantly changed between 0 and 5M KCl (data not shown). Whether this reflects a property of the protease or the substrate, is unclear.

This study indicates the usefulness of synthetic model substrates for characterization of biochemical properties of the HIV-1 protease. It shows that heptapeptides are sufficiently large to serve as substrates. Larger substrates up to 32 residues long have been shown to be acceptable substrates as well (13) whereas hexapeptides are no longer cleaved (14). It is not too surprising that heptapeptides are accepted because other aspartic proteases have similar abilities (15). It is assumed that a heptapeptide represents the minimal length to fit into the cleft formed by aspartic proteases and by the dimerized protease. The data presented here indicate that the sequence is not confined to X-Y-Pro-Z as suggested by Pearl and Taylor (16) but is of more general nature. The protease appears to recognize hydrophobic domains as was also pointed out by others (14).

We have previously pointed out the influence of conformation on cleavage of substrates by the HIV-1 protease which is unable to process a denatured gag polyprotein whereas a non-denatured gag is processed in vitro (6). The effect of conformation may be irrelevant with short synthetic peptide substrates because of their high degree of flexibility. Once placed inside the cleft of the protease it may be the enzyme which imposes the correct conformation on the small peptide.

Among the synthetic model substrates, we selected for analysis, was one which corresponded to a theoretically possible cleavage site between reverse transcriptase and RNase H which furthermore contained the PIV sequence to the right of the cleavage site. This peptide was, however, not cleaved which must be attributed to the amino acid glu to the left of the cleavage site. This result shows that a common denominator of cleavage sites is not the proline but a more general hydrophobicity only. From this result we predict that cleavage of the p66 reverse transcriptase/RNase H molecule into a p51 and a low molecular weight p15 molecule, which we have shown to be enzymatically active as RNase H (17), may be performed by a protease of non-viral origin.

Inhibition of the protease by synthetic peptides illustrates that it may be possible to design competitive inhibitors to prevent the action of the HIV-1 protease. One should, however, point out that the use of the synthetic substrates for design of HIV-1 inhibitors may not be sufficient. The effect of anti-viral drugs needs to be confirmed using the natural protease substrates.

Using a model substrate and the estimated protein content of the partially purified protease (maximally 1 μg/ml) as a basis for the estimation of the turnover number for substrate cleavage, we reach the conclusion that it cleaves at least 10 molecules per minute. This is about 10fold more efficient than the avian protease (14).

The retroviral proteases are less efficient than other aspartic proteases, e.g. pepsin A. This may reflect a biologically relevant property and an adjustment to the need for virus maturation. The number of cleavages required for viral protein processing can be estimated as follows: a virus particle requires approximately 60 min for budding, it contains about 2000 gag molecules which are cleaved three times each and approximately 60 reverse transcriptase molecules which are cut once. For autocatalysis two cuts are required which leads to about 6180 cuts per 60 min and about 100 cuts per min - 10 times more than determined here. It needs to be shown which in vitro parameters will improve the activity of the p9 protease by another factor of 10.

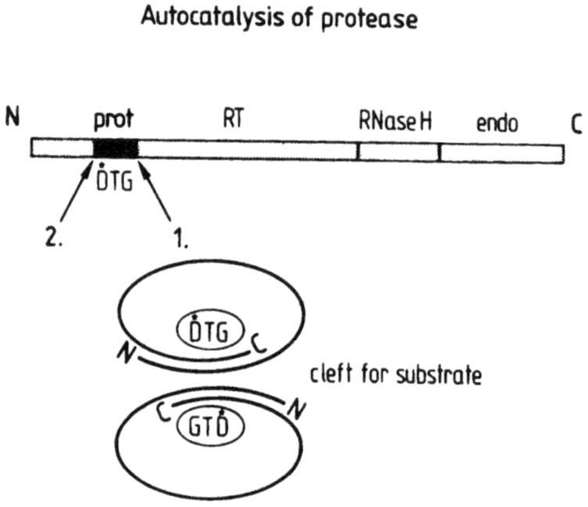

Fig. 3 Model of the HIV-1 protease.
The carboxy- and amino termini (C,N) of the protease monomer line up close to the active center D*TG. The two protease dimers form a cleft which allows substrates to be cleaved as well as autocatalytic processing of the protease molecules from the precursors. The carboxy-terminal end of the protease is cleaved first, the amino-terminal end is cut next (6), indicated by 1. and 2.

The autocatalytic cleavage of the p9 protease raises the question how it liberates itself from the p110 pol precursor. The model predicted for the HIV-1 protease (11) shows that both ends of the molecule fold back in such a way that they get close to the D*TG, the active center of the protease (Fig.3). We confirmed experimentally the necessity of the carboxy-terminus of the protease for its activity. Deletion of 17 carboxy-terminal amino acid residues results in complete loss of activity (6). It is also the carboxy-terminus of the p9 protease, which is cleaved first, as was determined by a kinetic analysis, from a larger precursor (6). Subsequently the amino-terminus gets cleaved (6). Therefore one can envision that the carboxy- and amino-terminal ends which line up parallel to each other close to the active center are part of the cleft between the dimerized protease. The uncleaved polyprotein precursors are probably folded alike and thus both termini are accepted as substrates and cleaved accordingly. Whether cleaveage works in cis or trans remains to be shown.

Acknowledgement

The technical assistance of S.Sukrow, S.Rabe and S.Grams is gratefully acknowledged. We thank I.Schallehn for secretarial work and E.Philippi for the technical drawings. This work was supported by the Bundesministerium für Forschung und Technologie and the Dr. Mildred Scheel-Stiftung.

REFERENCES

1. Barré-Sinoussi, F., Cherman, G.C., Rey, G., Nugeyre, M.T., Chamaret, S., Gruest, G., Dauguet, C., Axler-Blin, C., Vezinet-Brun, F., Rouzioux, C., Rozenbaum, W., and Montagnier, L. (1983) Science 220, 868-870.

2. Gallo, R.C., Salahuddin, S.Z., Popovic, M., Shearer, G.M., Kaplan, M., Haynes, B.F., Palker, T.G., Redfield, R., Oleske, G., Safai, B., White, G., Foster, P., and Markham, P. (1984) Science 224, 500-502.

3. Wong-Staal, F., and Gallo, R.C. (1985) Nature 317, 395-403.

4. Kohl, N.E., Emini, E.A., Schleif, W.A., Davis, J., Heimbuch, J.C., Dixon, R.A.F., Scolnick, E.J., and Sigal, J.S. (1988) Proc. Natl. Acad. Sci. USA 85, 4686-4690.

5. McCune, J.M., Rabin, L.B., Feinberg, M.B., Lieberman, M., Kosek, J.C., Reyes, G.R., and Weissman, I.L. (1988) Cell 53, 55-67.

6. Hansen, J., Billich, S., Schulze, T., Sukrow, S., and Moelling, K. (1988) EMBO J. 7, 1785-1791.

7. Debouck, C., Gorniak, J.G., Strickler, J.E., Meek, T.D., Metcalf, B.W., and Rosenberg, M. (1988) Proc. Natl. Acad. Sci. USA 84, 8903-8906.

8. Graves, M.C., Lim, J.J., Heimer, E.P., and Kramer, R.A. (1988) Proc. Natl. Acad. Sci. USA 85, 2449-2453.

9. Le Grice, S.F.J., Mills, J., and Mous, J. (1988) EMBO J. 7, 2547-2553.

10. Lillehoj, E.P., Salazar, F.H.R., Mervis, R.J., Raum, M.G., Chan, H.W., Ahmad, N., and Venkatesan, S. (1988) J. Virol. 62, 3053-3058.

11. Pearl, L.H., and Taylor, W.R. (1987) Nature 329, 351-354.

12. Kay, J., Jupp, R.A., Norey, C.G., Richards, A.D., Reid, W.A., Taggart, R.T., Samloff, J.M., and Dunn, B.M. (1987) In "Proteases: Potential role in health and disease" (Heidland,A. and Horl,W.H., eds.). Plenum Press.

13. Schneider, J., and Kent, S.B.H. (1988) Cell 54, 363-368.

14. Kotler, M., Katz, R.A., Danho, W., Leis, J., and Skalka, A.M. (1988) Proc. Natl. Acad. Sci. USA 85, 4185-4189.

15. Kay, J. (1985) In: "Aspartic Proteinases and Their Inhibitors" (V.Kostka, ed), Walter de Gruyter, Berlin, New York.

16. Pearl, L.H. and Taylor, W.R. (1987) Nature 328, 482.

17. Hansen, J., Schulze, T., Mellert, W., and Moelling, K. (1988) EMBO J. 7, 239-243.

PRELIMINARY CRYSTALLOGRAPHIC INVESTIGATION OF A PROTEASE FROM ROUS SARCOMA VIRUS

M. Miller, M. Jaskolski, M. Rao, A. Wlodawer

NCI-FCRF, BRI-Basic Research Program, P.O. Box B, Frederick, MD 21701

J. Leis

Department of Biochemistry, Case Western Reserve University, Cleveland, OH 44106

Introduction

The retroviral gag, pol, and env genes are translated as precursor polypeptides which are cleaved to produce the mature forms of the protein found in virions (1). Processing of the gag and pol gene protein precursors of avian sarcoma/leukosis viruses (ASLV), murine leukemia viruses (MuLV), and human immunodeficiency virus (HIV) requires a viral encoded protease PR[1] (1-4). The processing occurs late in the virus life cycle at or near the time of budding (1) and is essential for production of infectious virus particles. Viruses with mutations in the protease do not produce mature particles (2). This suggests that inhibitors of the viral protease could potentially serve as antiviral agents.

The retroviral proteases appear to be related to the cellular aspartic protease family both biochemically and by the presence of a conserved amino acid sequence, Asp-Thr/Ser-Gly, believed to be involved in the active site of the enzyme (5-7). The ASLV PR has been expressed in bacteria as intact PR and as a NC-PR fusion protein (9). Processing of the fusion protein in extracts is dependent upon an active PR since substitution of Ile for the conserved Asp-37 residue in the putative active site of the PR blocks the processing (9).

[1]Retrovirus protein nomenclature used in this discussion is as defined by Leis et al. (13).

Proteases of Retroviruses

The ASLV protease contains 124 amino acids. In contrast to mammalian viruses, the ASLV PR is encoded in the gag rather than the pol gene. Thus, in ASLV infected cells, the PR is synthesized as part of two precursor polypeptides, $Pr76^{gag}$ and $Pr180^{gag-pol}$, the former being expressed in concentrations 20 times that of the latter. As a consequence, the PR is present in high concentration in virus particles and can be readily purified.

The sites of cleavage in several retroviral precursor polypeptides have been deduced by alignment of the amino acid sequences at the C and N termini of the mature gag and pol proteins. Although some sequences occur frequently, no common features are apparent to account for the cleavage specificity. Studies utilizing small synthetic peptides as both model substrates and inhibitors for the PR have shown that the ASLV protease can specifically hydrolyze a peptide which extends 5 amino acids but not 3 amino acids in both directions from the α-IN cleavage site of the pol precursor polypeptide (8). A decamer representing the NC-PR cleavage site is also specifically cleaved (M. Kotler, A. Skalka, R. Katz, and J. Leis, unpublished observations). Substitutions of amino acids to either side of the peptide bond to be cleaved affect the ability of the peptide to serve as a substrate. Some substitutions produced peptides that are inhibitors rather than substrates.

Earlier reports have shown that the ASLV PR can process MuLV (10-12) and FeLV (5) gag polyprotein precursors to products that migrate like the authentic viral proteins. The ASLV PR can also hydrolyze two polypeptides that contain HIV cleavage sites (8). This apparent cross-specificity of the ASLV PR suggests that the three-dimensional structure of the protease will be important for engineering inhibitors with therapeutic value in the treatment of retroviral-induced diseases such as AIDS.

Purification of ASLV Protease

The ASLV protease was solubilized from virus by the chloroform-methanol-salt extraction procedure as previously described (4). The protein was purified as above with the following modifications. The lyophilized PR protein was dissolved in 10 ml of 10 mM β-mercaptoethanol, 0.1 mM EDTA, 10 mM Tris-HCl, pH 7.5, 0.1 M NaCl and chromatographed through a G75 Sephadex column (5 x 26 cm) equilibrated with the same buffer. Eight ml fractions were collected and protein was followed by absorbancy at 280 nm. The elution profile is shown in Fig 1A. Aliquots of the fractions containing between 20 - 30 μg of protein were analyzed by SDS-12 % polyacrylamide gel electrophoresis in Tris-bicine buffer as previously described (4) and a photograph of the gel is shown in Fig. 1B. The PR was found to elute as a peak with a shoulder between 43 and 59% column volume. When the G75 column was equilibrated with the above buffer except containing 1 M NaCl, the PR eluted as a symmetrical peak between 43 and 52% column volume (data not shown). The molecular weight of the PR under these conditions was estimated to be approximately 26 kD. Ovalbumin (43 kD), cytochrome C (12.5 kD) and soybean trypsin inhibitor (20 kD) were used to standardize the column. Since the molecular weight of the PR calculated from the known amino acid sequence is 13.5 kD (20), these results suggests that the PR isolated from virus is purified as a dimer. This would be consistent with predictions that the active form of the enzyme may be a dimer (7).

The fractions containing the PR were dialyzed overnight against 2 l of 10 mM β-mercaptoethanol and lyophilized. The PR was dissolved and assayed as described by Kotler et al., (8). The PR preparation purified under these conditions specifically cleaved both precursor polypeptides (4) and peptide substrates (8). Optimum activity was obtained between pH 5.5 - 6.0 at 37°C. The presence of high ionic strength during the assay stimulated the PR to cleave peptide substrates about 2-4 fold (M. Kotler, personal communication). The PR from AMV, RSV (Pr-C), and a variant virus, 95-C have been purified by this procedure and are enzymatically indistinguishable from one another. The PR used for the crystallization studies described below was derived form RSV (Pr-C) since it is isogenic.

Crystallization of RSV Protease

A 0.5 mM solution of the protease was used for screening for crystallization conditions, using the hanging-drop technique. Lyophilized

A

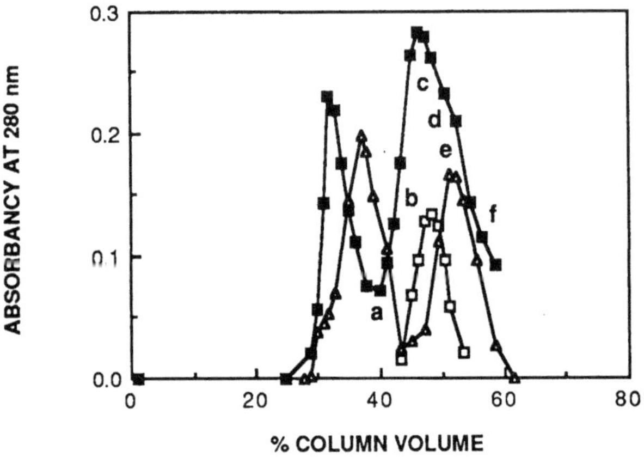

B

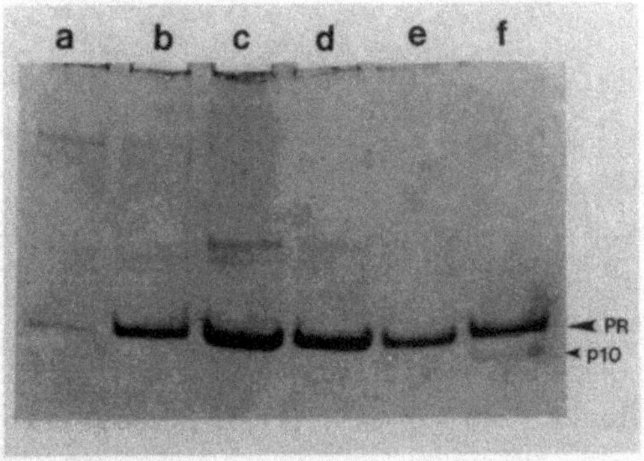

Fig. 1 **A**. Gel filtration of the ASLV PR. Partially purified PR preparations were subjected to gel filtration through a G75 Sephadex column at 4°C as described in text. Absorbancy at 280 nm for the PR preparation (■). Ovalbumin and cytochrome C (Δ) and soybean trypsin inhibitor (□) were used to standardize the gel filtration column in separate runs. **B**. SDS-polyacrylamide gel electrophoresis analysis of the PR preparation chromatographed through the gel filtration column. Fractions analyzed are indicated by the small letter in Fig. 1A. Proteins were detected by staining with coomassie blue. The migration position of purified PR and p10 proteins are indicated.

protein was dissolved in 20 mM sodium cacodylate buffer, pH 7, 1 mM β-mercaptoethanol. Clusters of microcrystals or very thin plates were obtained from a variety of conditions in the pH range 5 to 6. Hexagonal prisms, 0.2 mm thick and up to 0.4 mm long (Fig. 2), suitable for X-ray diffraction study were grown at room temperature under the following conditions: 4 μl of protein solution was mixed with an equal volume of reservoir solution containing 10% ammonium sulfate in 0.1 M sodium acetate buffer, pH 5.4, and 20 mM $MgCl_2$. Droplets were then sealed over 1 ml of this reservoir solution to which 20 μl of 0.1 M β-mercaptoethanol and 40 μl of dimethylformamide were added. Crystals usually appeared after three to five days and reached their maximum size of 0.4 mm x 0.2 mm x 0.2 mm within about one week.

Preliminary Crystallographic Investigation

Currently available crystals diffract beyond 2 Å resolution and are stable in the X-ray beam. The space group is $P3_121$, and the unit cell parameters are: a = b = 88.95 Å, c = 78.9 Å. X-ray data collected with a Nicolet Proportional Counter (14) were initially used to estimate unit-cell parameters, which were subsequently confirmed by precession photographs (Figure 3). A number of native data sets, as well as data for several putative heavy atom derivatives, were measured with an area detector. The best native data set consisted of 96,106 individual measurements of 20,694 unique reflections, which merged with a weighted R factor on intensities (14) of 6.9%. The number of reflections with significant intensity ($F>3\sigma(F)$) is 17,835 and the data are essentially complete to 2.1 Å.

Analysis of the amount of protein present in the unit cell (15) indicated the presence of either two (V_m=3.25) or three (V_m=2.17) molecules of PR in the crystallographic asymmetric unit. The latter possibility was excluded on the basis of estimates of crystal density in a gradient of aqueous ficoll (16). This result suggests the possibility of observing the protease as a dimer, its probable functional form (7). Moreover, the crystals were grown at the pH corresponding to its maximum enzymatic activity.

Fig 2. Crystals of the RSV (Pr-C) protease

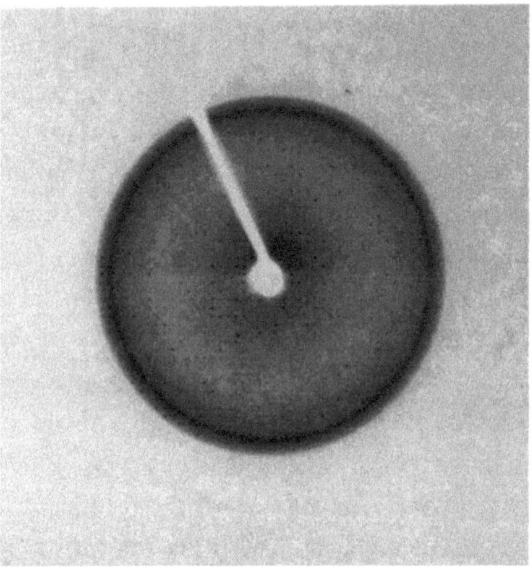

Fig 3. A precession photograph of the *Okl* layer. The precession angle μ is 14° and the resolution at the edge of the picture is 3.2 Å.

Heavy atom derivatives were prepared by soaking the crystals in a synthetic mother liquor containing appropriate salts in concentrations ranging from 1 to 6 mM. Several compounds of mercury, platinum, and uranium caused measurable changes in the scattered intensities. Soaking the crystals for 22 hours in a 2 mM solution of uranyl acetate yielded a single-site high-occupancy derivative, while several mercurials gave multisite, lower occupancy derivatives.

The strong anomalous signal of uranium was used to determine the enantiomorph and to place all derivatives at a common origin. Uranium phases in both enantiomorphic space groups and including anomalous scattering contribution were used to calculate difference Fourier maps for the mersalyl derivative. The heavy-atom peaks in the map calculated assuming $P3_121$ symmetry became ca 2.5 times lower when the Friedel pairs [F(+)/F(-)] of the uranyl derivative were exchanged. This indicates that the $P3_121$ is the correct space group. The major heavy-atom sites of all derivatives were clearly visible in the difference Patterson maps while the minor sites were located from several cross-Fourier maps. The heavy-atom parameters for each derivative were first refined separately yielding the agreement factors reported in Table 1. A combined three-dimensional refinement of all derivatives, including anomalous scattering for the uranyl derivative, gave a total figure of merit of 0.67 for 3 Å data. All Fourier calculations and lack-of-closure MIR refinements of the heavy-atom parameters were done using the PROTEIN program package (17).

Electron density maps based on the MIR phases revealed unambiguous boundaries of the protein regions and confirmed the presence of two molecules in the asymmetric unit. To improve the maps, the solvent flattening/phase extension technique of Wang (18) was applied. The starting phases for this procedure were obtained by phase combination of SIR and SAS data for the uranyl and methyl mercury acetate derivatives. The solvent flattening procedure also served to verify the space group assignment since the comparison of final statistics clearly indicated $P3_121$ as the correct space group: figure of merit 0.87 (0.85), map inversion R factor 0.184 (0.245), F_o/F_c correlation coefficient 0.977 (0.961) (values in parentheses are for $P3_221$).

Table 1. Centric-zone Refinement Statistics

	Resolution	R_{Cullis}	Figure of merit
Uranyl acetate 1 site	3Å	0.45	0.69
Mercury Acetate 4 sites	3Å	0.58	0.61
Mersalyl 4 sites	3Å	0.56	0.64
Methyl Mercury Acetate 3 sites	4Å	0.56	0.64

The solvent-flattened F_0 map calculated at 3 Å resolution was interpretable enough to permit a preliminary fitting of a polyalanine chain to the electron density. As a result, a model consisting of several stretches of polypeptides with a total of 214 residues was obtained. In this crude model two large β-barrel domains and two medium-length right-handed α-helices were recognized. However, the two independent molecules could not yet be unambiguously identified.

Least-squares refinement of the above model using the restrained refinement program of Hendrickson and Konnert (19) converged with an R factor of 0.38. The new phases obtained from the refined model are being currently used to produce new electron density maps from which the starting model will be extended and improved.

In parallel with the refinements, an extensive search has been carried out to locate the noncrystallographic symmetry between the two independent molecules in the asymmetric unit. This search has been done using both the electron density map and the 214-residue model. The results are consistent and indicate a two-fold axis perpendicular to $\mathbf{a}^*$, inclined to and shifted from the crystallographic diad along $\mathbf{b}$, as a possible dimer axis.

Acknowledgement

Research sponsored in part by the National Cancer Institute, DHHS, under contract N01-0-74101 with Bionetics Research, Inc. J. L. is supported in part by Public Health Service Grant CA 38046 from the National Cancer Institute and a pilot research grant from the American Cancer Society, Ohio, Cuyahoga county division. The contents of this publication do not necessarily reflect the views or policies of the Department of Health and Human Services, nor does mention of trade names, commercial products, or organizations imply endorsement by the U.S. Government.

References

1. Dickson, C., R. Eisenman, H. Fan, E. Hunter, and N. Teich, 1984. In: RNA Tumor Viruses, Vol. 1 (Weiss, R., Teich, N., Varmus, H., and Coffin, J., eds.). Cold Spring Harbor Laboratory, Cold Spring Harbor, N.Y. pp. 513-648.

2. Voynow, S. and J. Coffin. 1985. J. Virol. **55,** 79-85.

3. Katoh, I., Y. Yoshiyuki, A. Rein, M. Shihuya, T. Odaka, T., and Oroszlan, S. 1985. Virology **145**, 280-292.

4. Alexander, F., Leis, J., Soltis, D., Crowl, R., Danho, W., Poonian, M., Pan, Y-C., and Skalka, A. 1987. J. Virol. **61**, 534-542.

5. Toh, H., Ono, M., Saigo, K., and Miyata, T. 1985. Nature **315**, 691.

6. Miller, R. 1987. Science **236**, 722-725.

7. Pearl, L. and Taylor, W. 1987. Nature **329**, 351-354.

8. Kotler, M., Katz, R., Danho, W., Leis, J. and Skalka, A. 1988. Proc. Natl. Acad. Sci. U.S.A., **85**, 4185-4189.

9. Kotler, M., Katz, R., and Skalka, A.M. 1988. J. Virol. **62**, 2696-2700.

10. Demsey, A., Collins, F., and Kanaka, D. 1980. J. Virol. **36**, 872-876.

11. Pepinsky, R. 1983. J. Biol. Chem. **258**, 11229-11235.

12. Yoshinaka, Y., and Luftig, R. 1981. Virology **111**, 239-250.

13. Leis, J., Baltimore, D., Bishop, J., Coffine, J., Fleissner, E., Goff, S., Oroszlan, S., Robinson, H., Skalka, A., Temin, H., and Vogt, V. 1988. J. Virol. **62**, 1808-1809.

14. Howard, A.J., Gilliland, G.L., Finzel, B.C., Poulos, T.L., Ohlendorf, D.H. and Salemme, F.R. 1987. J. Appl. Crystallorgr. **20,** 383-387.

15. Matthews, B.W. 1968. J. Mol. Biol. **33**, 491-497.

16. Westbrook, E.M. 1985. In: Methods in Enzymology, Vol. 114, (Wyckoff, H.W., Hirs, C.H. and Timasheff, S.N., eds.). Academic Press, Inc. pp. 187-196.

17. Steigemann, W. 1972. Ph. D. Thesis, Technical University, Munich, FRG

18. Wang, B. C. 1985. In: Methods in Enzymology, Vol. 115, (Wyckoff, H.W., Hirs, C.H. and Timasheff, S.N., eds.). Academic Press, Inc. pp. 90-112.

19. Hendrickson, W. A., Konnert, J. H. 1981. In: Biomolecular Structure, Conformation, Function and Evolution (Srinivasan, R., ed.), Vol. 1, Pergamon Press, Oxford, England, pp. 43-57.

20. Schwartz, D., Tizard, R., Gilbert, W. 1983. Cell **32**, 853-869.

STRUCTURE AND EVOLUTION OF RETROVIRAL PROTEINASES

Laurence H. Pearl

Laboratory of Molecular Biology, Department of Crystallography, Birkbeck College, University of London, Malet Street, London WC1E 7HX.
and
Laboratory of Virology, Section of Cell and Molecular Biology, Institute of Cancer Research, Chester Beatty Laboratories, Fulham Road, London SW3 6JB.

Introduction

In 1985, Toh *et al* (1,2) observed that a cluster of amino acid residues DT(S)G[1] in the sequences of several transposable elements and retroviruses, were homologous to a sequence in the p15^{gag} proteinase of Rous sarcoma virus and to sequences forming the active sites of several aspartic proteinases, and suggested that the retroviral proteinases may possess a similar activity to the aspartic proteinases.

Retroviral Proteinase Sequences

Since 1985, nucleotide sequences for over twenty retroviruses and retrotransposons have become available, including the

---

[1] Single letter amino acid codes are used throughout this article. In addition, '•' signifies any hydrophobic residue. Where two alternative amino acids can occur, the second is given in parentheses after the first.

Proteases of Retroviruses

Retroviral Proteinases

```
              5    10   15   20   25   30   35   40   45   50   55   60
              |    |    |    |    |    |    |    |    |    |    |    |
FeLV      ----KIRRVRARTPPPEPRITL---RIGGQPVTFLVDTGAQHSVLTRPDGPLSD--RTAL---
M-RadLV   -------GGQGQEPPPEPRITL---KVGGQPVTFLVDTGAQHSVLTQNPGPLSD--RSAW---
Mo-MLV    --TLDDQGGQGQDPPPEPRITL---KVGGQPVTFLVDTGAQHSVLTQNPGPLSD--KSAW---
AKV-MLV   --TLDDQGGQGQEPPPEPRITL---TVGGQPVTFLVDTGAQHSVLTQNPGPLSD--RSAW---
BaEV      --EDSEQGCQGSGAPPEPRLTL---SVGGHPTTFLVDTGAQHSVLTKANGPLSS--RTSW---

BLV       -LSIPLARSRPSVAVYLSGPWL---QPSQNQALMLVDTGAENTVLPQNWLVR-DYPRIPA---
STLV      -LPIIPLDPARRPLIKAEVNTQ---TSHPMIIEALLDTGADMTVLPMALFS-DDIPLKNT---
HTLV-I    -LPVIPLDPARRPVIKAQVDTQ---TSHPKTIEALLDTGADMTVLPIALFS-SNTPLKNT---
HTLV-II   -LPLIPLRQQQQPILGVRISVM---GQTPQPTQALLDTGADLTVIPQTLVP-GPVKLHDT---

HIV1bru   --PQITLWQRPLVTIKIG----------GQLKEALLDTGADDTVLEEMSLP-GRW-KPKM---
HIV1sf2   --PQITLWQRPLVTIRIG----------GQLKEALLDTGADDTVLEEMNLP-GKW-KPKM---
HIV1mal   --PQITLWQRPVVTVRVG----------GQLKEALLDTGADDTVLEEINLP-GKW-KPKM---
HIV1eli   --PQITLWQRPLVAIKIG----------GQLKEALLDTGADDTVLEEMNLP-GKW-KPKM---
SIVagm    --PLWR---RPIKTVYIE----------GVPIKALLDTGADDTIIKENDLQSGPW-RPKI---
HIV2rod   --PQFSLWKRPVVTAYIE----------GQPVEVLLDTGADDSIVAGIELG-NNY-SPKI---
SIVmac    --PQFSLWRRPVVTAHIE----------GQPVEVLLDTGADDSIVTGIELG-PHY-TPKI---
VLV       --PYVVTEAPPKIEIKVG----------TRWKKLLVDTGADKTIVTSHDMS-GIPKGRII---
EqIAV     --VTYNLEKRPTTIVLIN----------DTPLNVLLDTGADTSVLTTAHYNRLKYRGRKYQ--

MMTV      -WVQEISDSRPMLHIYLN----------GRRFLGLLDTGADKTCIAGRDWP-ANWPIHQTESS
SRV-1     -WVQPITCQKPSLTLWLD----------DKMFTGLIDTGADVTIIKLEDWP-PNWPITDTLTN
MPMV      -WVQPITCQKPSLTLWLD----------DKMFTGLIDTGADVTIIKLEDWP-PNWPITATLTN
MuIAP     -SSGNDSAYLVVSLNDRPKLR---LKINGKEFEGILDTGADKSIISTHWWP-KAWPTTESSHS
IAP-H18   -KKDCQAPERTRESKLCYRCG---KGYHRASECGIMDSGADKSIISLHWWP-KSWPTVVSSHS
HERV      -PTGKAAYWASQVSENRPVCK---AIIQGKQFEGLVDTGADVSIIALNQWP-KNWPKQKAVTG

RSV       LAMTMEHKDRPLVRVILTNTGSHPVKQRSVYITALLDSGADITIISEEDWP-TDWPVMEAANP
AMV       LAMTMEHKDRPLVRVILTNTGSHPVKQRSVYITALLDSGADITIISEEDWP-ADRPVMEAANP

HFV       TVTQSATSSTDESSSAVTAASG---RIKGTKLLAWHDSGATITCIPESFLE-DEQPIKKTLIK
```

```
          65   70   75   80   85   90   95  100  105  110  115  120  125  130  135  140
          |    |    |    |    |    |    |    |    |    |    |    |    |    |    |    |
FeLV      -VQGATGSKNYRWT----TDRRVQL--ATGKVTH-SFLYVPECP-YPLLGRDLLTKLKAQIHFTGEGANVVGPRGLPLQVL
M-RadLV   -VQGATGGKRYRWT----TDRKVHL--ATGKVTH-SFLHVPDCP-YPLLGRDLLTKLKAQIHFKGSGAQIVGPMGQPLQVL
Mo-MLV    -VQGATGGKRYRWT----TDRKVHL--ATGKVTH-SFLHVPDCP-YPLLGRDLLTKLKAQIHFEGSGAQVMGPMGQPLQVL
AKV-MLV   -VQGATGGKRYRWT----TDRKVHL--ATGKVTH-SFLHVPDCP-YPLLGRDLLTKLKAQIHFEGSGAQVVGPKGQPLQVL
BaEV      -VQGATGRKMHKWTN----RRTVNL--GQGMVTH-SFLVVPECP-YPLLGRDLLTKLGAQIHFSEAGAQVLDRDGQPIQIL

BLV       AVLGAGGVSRN-RYNWL-QGPLTLALKPEGPFITIPKILVDTFDKWQILGRDVLSRLQASISIPEEVRPRMVG
STLV      SVLGAGGQTQD-HFKLT-SLPVLIRLPFRTTPIVLTSCLVDTKNNWAIIGRDALQQCQGVLYLPEAKGPRTIL
HTLV-I    SVLGAGGQTQD-HFKLT-SLPVLIRLPFRTTPIVLTSCLVDTKNNWAIIGRDALQQCQGVLYLPEAKGPRVIL
HTLV-II   LILGASGQTNT-QFKLL-QTPLHIFLPFRRSPVILSSCLLDTHNKWTIIGRDALQQCQGLLYLPDDRSPHQLL

HIV1bru   -IGGIGGFIKVRQYD-----QILIEICGHK---AIGTVLVGPTP-VNIIGRNLLTQIGCTLNF
HIV1sf2   -IGGIGGFIKVRQYD-----QIPVEICGHK---AIGTVLVGPTP-VNIIGRNLLTQIGCTLNF
HIV1mal   -IGGIGGFIKVRQYD-----QILIEICGKK---AIGTILVGPTP-VNIIGRNMLTQIGCTLNF
HIV1eli   -IGGIGGFIKVRQYD-----QIPIEICGQK---AIGTVLVGPTP-VNIIGRNLLTQIGCTLNF
SIVagm    -IGGIGGGLNVKEYN---DREVKIEDKILR-----GTILLGATP-INIIGRNLLAPAVPRLVM
HIV2rod   -VGGIGGFINTKGYK-----NVEIEVLNKK---VRATIMTGDTP-INIFGRNILTALGMSLNL
SIVmac    -VGGIGGFINTKEYK-----NVEIEVLGKR---IKGTIMTGDTP-INIFGRNLLTALGMSLNL
VLV       -LQGIGGIIEGEKWE-----QVHLQYKDKM---IKGTIVVATSP-VEVLGRDNMRELGIGLIM
EqIAV     GTGIIGVGGNVETFS----TPVTIKKKGRH---IKTRMLVADIP-VTILGRDILQDLGAKLVL

MMTV      -LQGLGMACGVARSS---QPLRWQHEDKS---GIIHPFVIPTLP-FTLWGRDIMKDIKVRLMTDSPDDSQDL
SRV-1     -LRGIGQSNNPKQSS---KYLTWRDKENNS--GLIKPFVIPNLP-VNLWGRDLLSQMKIMMCSPSDIVTAQM
MPMV      -LRGIGQSNNPKQSS---KYLTWRDKENNS--GLIKPFVIPNLP-VNLWGRDLLSQMKIMMCSPNDIVTAQM
MuIAP     -EQGLGYQSCPTASS---VALTWESSEGQQ--GKFIPYVLP-LP-VNLWGRDIMQHLGLILSNENAPSGGYS
IAP-H18   -LQGLGYQSSPAISA---SALTWRDAEGKQ--GCFTPYVLP-LP-VNLWGRDVLQAMGMTLTNEYSPQASAI
HERV      -LVGIGTASEVYQSM---EILHCLGPDNQE--STVQPMITS-IP-LNLWGRDLLQQWGAEITMPARLYSPTS

RSV       QIHGIGGGIPMRKSRDMIELGVINRDGSLE--RPLLLFPAVAMVRGSILGRDCLQGLGLRLTNL
AMV       QIHGIGGGIPMRKSRDMIEVGVINRDGSLE--RPLLLFPAVAMVRGSILGRDCLQGLGLRLTNL

HFV       TIHGEKQQNVYYVTFKVKGRKVEAEVIASP----YEYILLSPTD-VPWLTQQPLQLTILVPLQ
```

Figure 1.

Alignment of proteinase sequences from : feline, murine radiation, Moloney and AKV murine leukemia viruses (3-6), baboon endogenous virus (7), bovine, simian and human types 1 and 2 leukemia viruses (8-11), HIV-1 isolates <u>bru</u>, <u>sf2</u>, <u>mal</u> and <u>eli</u> (12-14), SIV from african green monkey (15), HIV-2 isolate <u>rod</u> (16), SIV from macaque (17), visna lentivirus (18), equine infectious anaemia virus (19), mouse mammary tumour virus (20), simian retrovirus 1 (21), Mason-Pfizer monkey virus (22), murine and hamster intracisternal A-particles (23,24), human endogenous retrovirus (25), Rous sarcoma virus (26), avian myeloblastosis virus complex (27), human foamy (spumaretro) virus (28).

sequences of several strains of the human immunodeficiency viruses (HIV-1, and HIV-2) which are believed to be the causal agents of AIDS. All the available sequences of replication competent retroviruses show regions with homology to the cluster of amino acids first observed by Toh *et al.* Figure 1 shows an alignment of the currently available retroviral proteinase coding sequences. In addition to the DTG cluster, several other regions of conservation are evident in this alignment: a sequence ••GRD(N) occurring some 60 or more residues after the DT(S)G, and a glycine rich region occurring between the DT(S)G and the ••GRD(N) sequences.

Apart from these three regions of conservation, the overall homology between all of these retroviral proteinase sequences is generally low. The sequence from the human foamy (spumaretro) virus shows by far the greatest difference in sequence to all the other viruses, and entirely lacks the otherwise totally conserved ••GRD(N) sequence towards the carboxy terminal. Amongst the human and simian immunodeficiency viruses, the close homology amongst the different HIV-1 isolates and between HIV-2 and SIV_{MAC}, is evident, while the sequence from SIV_{AGM}, shows a similar homology to the HIV-2/SIV_{MAC} group and the HIV-1 group. Sequence homologies based upon this alignment reflect the conventional classification of retroviruses.

Homology with Aspartic Proteinases

Crystallographic studies of aspartic proteinases (29,30,31,32) show them to consist of two distinct domains divided by an extensive cleft. The polypeptide chains in the two domains have similar topological structure, and show some homology between the amino acid sequences in each domain, particularly around the highly conserved ••DTG sequences (33). The conserved aspartic acids in the ••DTG from each domain are the

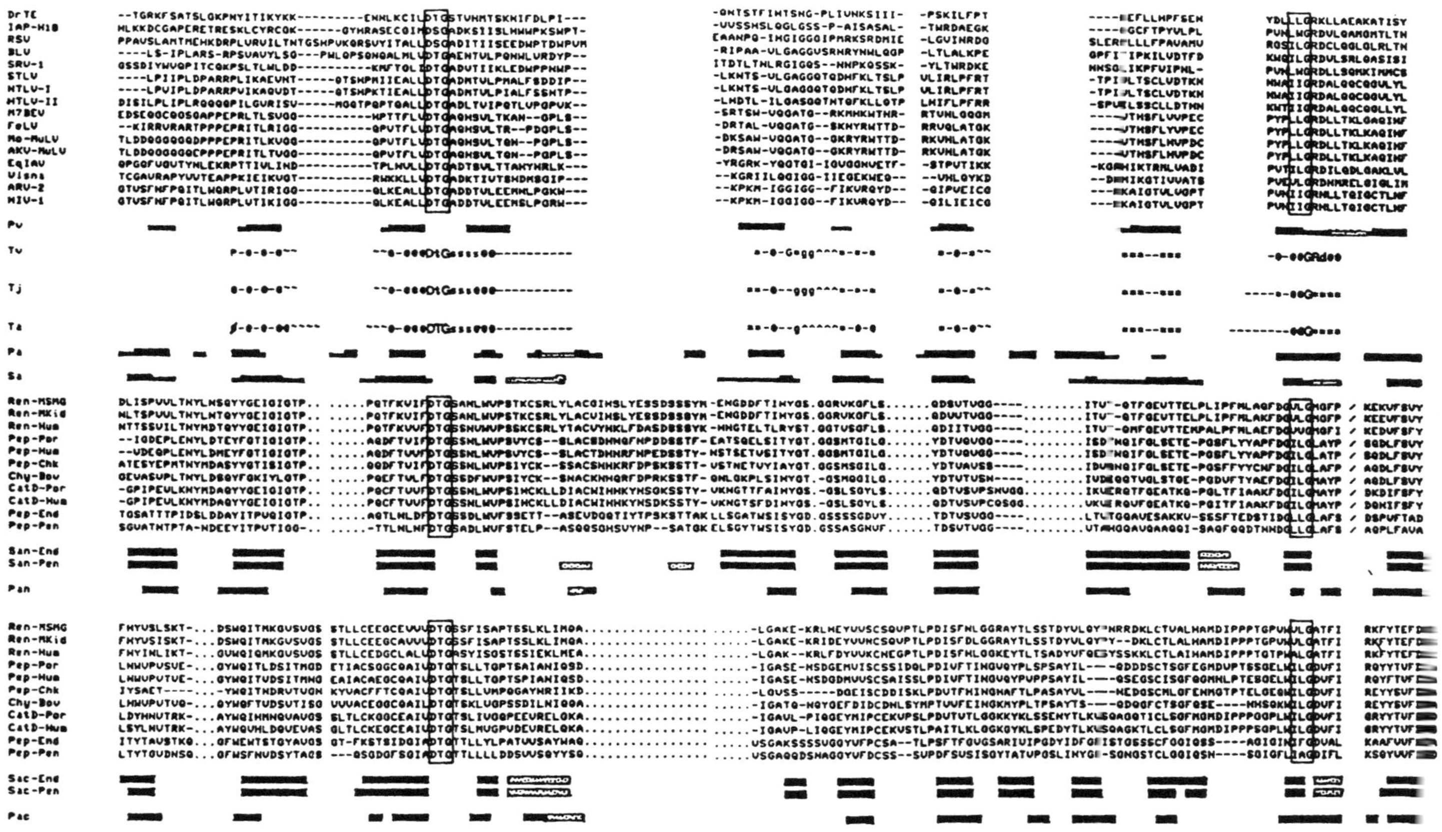

Figure 2. Reproduced from *Nature*. For a full explanation see reference 36.

residues known to be involved in the hydrolysis of substrate peptide bonds catalysed by these enzymes (34).

While the presence of the ••DTG pattern in the retroviral proteinase and aspartic proteinase sequences suggests a structural relationship between these enzymes, the retroviral proteinases only contain one ••DTG sequence and do not exceed 130 residues in length, whereas the aspartic proteinases contain two ••DTG sequences and are all longer than 300 residues. Clearly therefore, the retroviral sequences can only correspond to a single domain of an aspartic proteinase. Conventional alignment techniques are incapable of satisfactory alignment of individual sequences where the homology is low, and restricted to small clusters of conserved residues. Using 'template' and secondary structure alignment techniques (35) a consensus alignment can be achieved between the retroviral proteinases and the separate domains of the aspartic proteinases (36). In this alignment (Figure 2), the major homologies between the retroviral sequences and the aspartic proteinases domains occur at the ••DTG and at the retroviral ••GRD(N) sequence which corresponds to a highly conserved ••G sequence found in both domains of the aspartic proteinases. In fact this ••G sequence, although distant from the active site ••DTG in the primary structure, is in very close proximity in the tertiary fold of the aspartic proteinase domains, and its conservation is important for the correct structure of the active site. The conservation of a homologous sequence at a similar position in the retroviral sequences suggests that it may play a similar role in those structures. A third region of homology occurs between a glycine rich segment occurring between the ••DTG and ••GRD(N) sequences in the retroviral proteinases, and the glycine rich β-hairpin 'flap' which is lies over the active site in the amino-terminal domains of the aspartic proteinases. The corresponding sequence in the retroviral enzymes may serve a similar role.

Crystal structure determinations of aspartic proteinases show them to be predominantly composed of β-sheet with < 10% α-helix. Generally, secondary structure prediction of an individual sequence is of limited use, and in the absence of a crystal structure for a related protein, its accuracy cannot be assessed. When a known structure is available, the reliability of the prediction technique can be measured, for that protein, by comparing the predicted secondary structure with that actually observed in the crystal structure of the protein. Even then, the prediction will tend to be influenced by the detailed features of the individual sequence. By taking a consensus of the individually predicted secondary structures for each sequence in an alignment, we produce a prediction that reflects the propensity to a particular type of secondary structure that may be conserved between the related sequences at each position, and may be used as a model of the secondary structure of a class of proteins.

Consensus secondary structure prediction on aligned aspartic proteinase domains predicts the positions of the β-strands actually observed in crystal structures with reasonable accuracy. Consensus prediction of the aligned sequences of the retroviral proteinases, generates a predominantly β-sheet structure, which corresponds well in position with the individual domains of aspartic proteinases. Preliminary low-resolution X-ray crystallography of the proteinase from Rous sarcoma virus, does appear to show a structure composed predominantly of β-sheet (Wlodawer,A. and Leis,J, personal communication).

Retroviral Proteinase Structure and Active Site Formation

Analysis of the retroviral proteinase sequences suggests that they are β-sheet structures with a similar number of strands to an aspartic proteinase domain. The most conserved regions in the retroviral sequences are also those which show the most homology with the aspartic proteinase domains, and correspond to the sequences which form the active site in aspartic proteinases. Imposing the common topology of the aspartic proteinase domains on the sequence of a retroviral proteinase, enables the construction a hypothetical three-dimensional model of a retroviral proteinase. The topology and three dimensional structure of the the model of the HIV-1 proteinase (SF2 isolate sequence) is shown in Figure 3. The topology is a 'trimmed' version of the aspartic proteinase domain fold, in which the central core of β-strands is very similar, but in which the external loops have been truncated. The 99 amino-acids of the HIV-1 sequence is just sufficient to form the central core. The highly conserved ••DTG and ••GRD(N) sequences come together to form a 'psi' structure identical with that found in each domain of aspartic proteinases.

The active sites of aspartic proteinases are formed by the conjunction of two DTG sequences, one from each of the two domains of the protein. The overall structural similarity between the two domains is restricted to comparable topology and some sequence homology, but in the active site, the DTG residues from each domain are related to each other by a virtually perfect twofold axis (37), and play key roles in the structure and function of the active site.The dimeric structure formed by the two aspartate residues serves to bind and activate a water molecule which is probably the attacking nucleophile in the hydrolysis of peptides by these enzymes (38,39). The threonine residues stabilise the active site structure by a dimeric side-chain to main-chain interaction (the 'fireman's grip'), where the side chain hydroxyl of each domain's threonine, hydrogen bonds to the main chain containing the threonine from the other domain. In some of the intracisternal A-particles, avian sarcoma viruses and the human foamy virus, this threonine is replaced by a serine which is the only amino acid other than threonine to possess the alcohol group necessary for this interaction. Recently, an

Figure 3.

a) Predicted topology of HIV-1 proteinase.
b) Stereo diagram of predicted HIV-1 proteinase monomer.

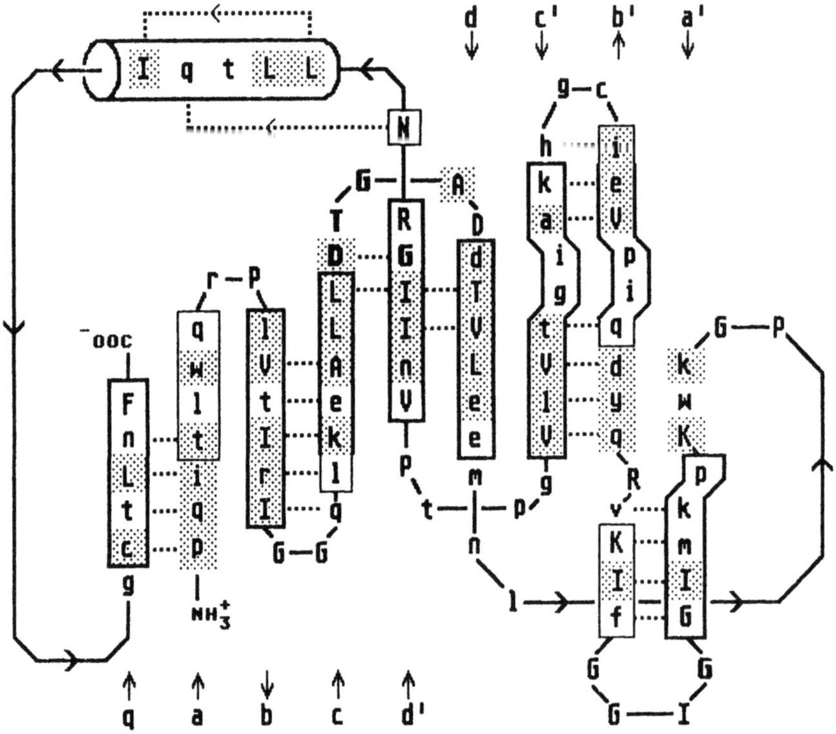

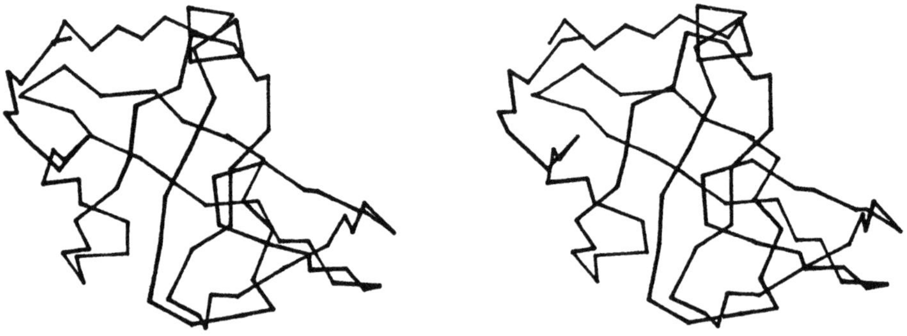

aspartic proteinase coded by the BAR1 gene of *S. cerevisiae*, has also been found to contain a DSG sequence in the active site of its carboxy-terminal domain (40). The glycine residues form tight bends and direct their peptide carbonyl groups upwards from the base of the substrate binding cleft. Crystal structures of several complexes between aspartic proteinases and inhibitors (41), show that these carbonyl groups, which are also related by the twofold axis, interact with the main chain of a substrate peptide and position it over the catalytic centre formed by the two aspartates.

The retroviral proteinase sequences only possess a single DTG sequence and could not function singly as aspartic proteinases. However if the retroviral proteins were capable of dimerising so that the DTG residues from two separate molecules could form the same interactions as in the active site structure described above, then they could function as dimeric single-domain aspartic proteinases[2]. An increasing body of biochemical evidence (in this volume) suggests that retroviral proteinases do indeed dimerise, but whether they actually form an active site structure similar to the aspartic proteinases is not yet clear.

---

[2] The amino acid sequence translated from the cloned nucleotide sequence of an HIV-2 isolate (HIV2NIHZ) shows a mutation from DTG to DRG. It is extremely unlikely that such a sequence would be able to form the required active site structure, and would not be expected to be competent as a proteinase. It is interesting to note that the viral clone from which this sequence was derived is non-infectious and cannot produce competent virions (G. Franchini, personal communication).

Evolution of Retroviral Proteinases

The amino and carboxy terminal domains of the aspartic proteinases have a similar topology and show a degree of sequence homology to each other, particularly in the regions of each domain that contribute to the active site. These observations led to the suggestion that the two domain aspartic proteinases may have arisen as a result of tandem duplication and fusion of an ancestral domain (33) (Figure 4). Aspartic proteinases have been identified in all phyla of eukaryotes, but no prokaryotic example has yet been identified. Comparison of aspartic proteinases from organisms as far apart as yeast and humans, shows a much higher homology between the same domains of enzymes from different species, than between different domains from the same species, suggesting that the duplication must have occurred prior to the speciation of an ancestral eukaryote.

Retroviruses are thought to have arisen from retro-transposable elements which 'escaped' from their host genome (42). The retroviral proteinases may be 'fossil' examples of the ancestral single domain, possibly active as a dimer, which gave rise to the aspartic proteinases. If this is the case, then the ancestral retrovirus escaped prior to the gene duplication event which gave rise to the modern aspartic proteinases (Figure 5).

Acknowledgement

This work is supported by the Medical Research Council AIDS Directed Research Programme.

Figure 4.

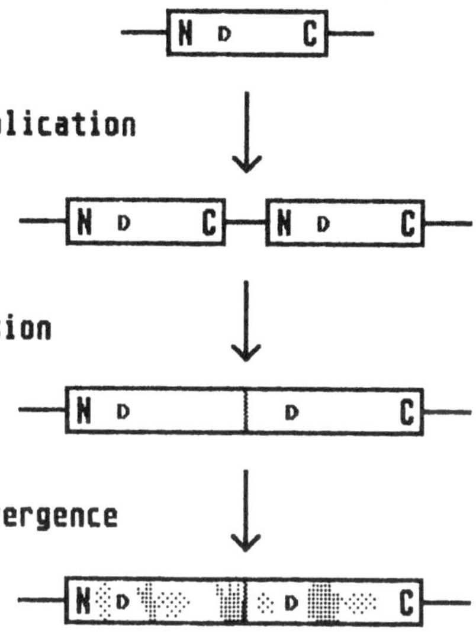

Figure 5.

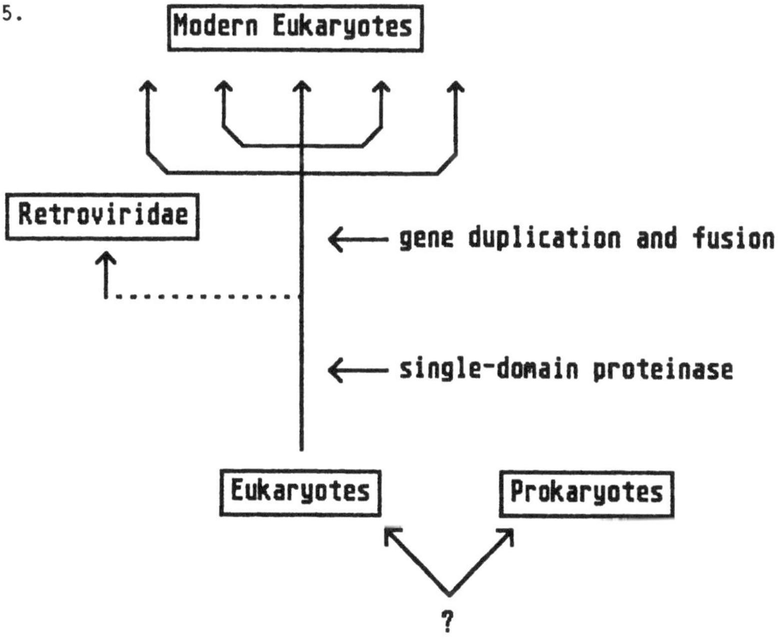

References

1. Toh, H., Ono, M., Saigo, K., Miyata, T. (1985), *Nature*, 315, 691.

2. Toh, H., Kikuno, R., Hayashida, H., Miyata, T., Kugimiya, W., Inouye, S., Yuki, S., Saigo, K. (1985), *EMBO J.*, 4, 1267-1272.

3. Laprevotte, I., Hampe, A., Sherr, C.J., Galibert, F. (1984), *J.Virol.*, 50, 884-894.

4. Merregaert, J., Janowski, M., Reddy, E.P. (1987), *Virology*, 158, 88-102.

5. Shinnick, T.M., Lerner, R.A., Sutcliffe, J.G. (1981), *Nature*, 293, 543-548.

6. Herr, W. (1984), *J.Virol.*, 49, 471-478.

7. Tamura, T-A. (1983), *J.Virol.*, 47, 137-145.

8. Rice, N.R., Stephens, R.M., Burny, A., Gilden, R.V. (1985), *Virology*, 142, 357-377.

9. Inoue, J-I., Watanabe, T., Sato, M., Oda, A., Toyoshima, K., Yoshida, M., Seiki, M. (1986), *Virology*, 150, 187-195.

10. Seiki, M., Hattori, S., Hirayama, Y., Yoshida, M. (1983), *Proc.Natl.Acad.Sci. USA*, 80, 3618-3622.

11. Shimotohno, K., Takahashi, Y., Shimizu, N., Gojobori, T., Golde, D.W., Chen, I.S.Y., Miwa, M., Sugimura, T. (1985), *Proc.Natl.Acad.Sci. USA*, 82, 3101-3105.

12. Wain-Hobson, S., Sonigo, P., Danos, O., Cole, S., Alizon, M. (1985), *Cell*, 40, 9-18.

13. Sanchez-Pescador, R., Power, M.D., Barr, P.J., Steimer, K.S., Stempien, M.M., Brown-Shimer, S.L., Gee, W.W., Renard, A., Randolph, A., Levy, J.A., Dina, D., Luciw, P.A. (1985), *Science*, 227, 484-492

14. Alizon, M., Wain-Hobson, S., Montagnier, L., Sonigo, P. (1986), *Cell*, 46, 63-74.

15. Fukasawa, M., Miura, T., Hasegawa, A., Morikawa, S., Tsujimoto, H., Keizaburo, M., Kitamura, T., Hayami, M. (1988), *Nature*, 333, 457-461.

16. Guyader, M., Emerman, M., Sonigo, P., Clavel, F., Montagnier, L., Alizon, M. (1987), *Nature*, 326, 662-669.

17. Chakrabarti, L., Guyader, M., Alizon, M., Daniel, M.D., Desrosiers, R.C., Tiollais, P., Sonigo, P. (1987), *Nature*, 328, 543-547.

18. Sonigo, P., Alizon, M., Staskus, K., Klatzmann, D., Cole, S., Danos, O., Retzel, E., Tiollais, P., Haase, A., Wain-Hobson, S. (1985), *Cell,* 42, 369-382.

19. Stephens, R.M., Casey, J.W., Rice, N.R. (1986), *Science,* 231, 589-594.

20. Moore, R., Dixon, M., Smith, R., Peters, G., Dickson, C. (1987), *J. Virol.,* 61, 480-490.

21. Power, M.D., Marx, P.A., Bryant, M.L., Gardner, M.B., Barr, P.J., Luciw, P.A. (1986), *Science,* 231, 1567-1572.

22. Sonigo, P., Barker, C., Hunter, E., Wain-Hobson, S. (1986), *Cell,* 45, 375-385.

23. Mietz, J.A., Grossman, Z., Lueders, K.K., Kuff, E.L. (1987), *J. Virol.,* 61, 3020-3029.

24. Ono, M., Toh, H., Miyata, T., Awaya, T. (1985), *J.Virol.,* 55, 387-394.

25. Ono, M., Yasunaga, T., Miyata, T., Ushikubo, H. (1986), *J.Virol.,* 60, 589-598.

26. Schwartz,D.E., Tizard,R., Gilbert,W. (1983), *Cell,* 32, 853-869.

27. Sauer, R.T, Allen, D.W., Niall, H.D. (1981), *Biochemistry,* 20, 3784-3791.

28. Maurer, B., Bannert, H., Darai, G., Flügel, R.M. (1988), *J.Virol.,* 62, 1590-1597.

29. Andreeva, N.S., Fedorov, A.A., Gutschina, E.E., Riskulov, R.R., Shutskever, N.E., Safro, M.G. (1978), *Mol. Biol. (Moscow),* 12, 922-936.

30. James,M.N.G., Sielecki,A. (1983), *J. Mol. Biol.,* 163, 299-361.

31. Blundell, T.L., Jenkins, J.A., Pearl, L.H., Sewell, B.T., Pedersen V.B. (1985), in *Aspartic Proteinases and their Inhibitors* (ed. Kostka, V.) 151-161, Walter de Gruyter and Co., Berlin.

32. Bott, R., Subramanian, E., Davies, D.R. (1982), *Biochemistry,* 21, 6956-6962.

33. Tang, J., James, M.N.G., Hsu, I-N., Jenkins, J.A., Blundell, T.L. (1977), *Nature,* 271, 618-621.

34. Chen, K.C.S., Tang, J. (1972), *J. Mol. Biol.,* 247, 2566-2574.

35. Taylor, W.R. (1986), *J. Mol. Biol.,* 188, 233-258.

36. Pearl, L.H., Taylor, W.R. (1987), *Nature,* 329, 351-354.

37. Pearl, L.H., Blundell, T.L. (1984), *FEBS Lett.,* 174, 96-101.

38. Pearl, L.H. (1985), in *Aspartic Proteinases and their Inhibitors* (ed. Kostka, V.) 189-195, Walter de Gruyter and Co., Berlin.

39. Pearl, L.H. (1987), *FEBS Lett.*, 214, 8-12.

40. Mackay, V.L., Welch, S.K., Insley, M.T., Manney, T.R., Holly, J., Saari, G.C., Parker, M.L. (1988), *Proc. Natl. Acad. Sci. USA*, 85, 55-59.

41. Foundling, S.I., Cooper, J., Watson, F.E., Cleasby, A., Pearl, L.H., Sibanda, B.L., Hemmings, A., Wood, S.P., Blundell, T.L., Valler, M.J., Norey, C.J., Kay, J., Boger, J., Dunn, B.M., Leckie, B., Jones, D.M., Atrash, D., Hallett, A., Szelke, M. (1987), *Nature*, 327, 349-352.

42. Temin, H.M. (1980), *Cell*, 21, 599-600.

AUTHOR INDEX

ABBREVIATIONS

AIDS	acquired immunodeficiency syndrome
AMV	avian myeloblastosis virus
ASLV	avian sarcoma/leukosis viruses
BaEV	baboon endogeneous leukemia virus
BLV	bovine leukemia virus
BPTI	basic pancreatic trypsin inhibitor
BSA	bovine serum albumin
CA	capsid protein
DAN	diazoacetyl D,L-norleucine methyl ester
DMS	dimethylsuberimidate
DMSO	dimethyl sulfoxide
DNM	1-deoxynojirimycin
Dns	5-dimethylaminonaphthalene-1-sulfonyl
DSS	disuccinyl suberate
DTT	dithiothreitol
EDTA	ethylenediamine tetraacetic acid
Endo H	endoglycosidase H
EPNP	1,2-epoxy-3-(p-nitrophenoxy)propane
EqIAV	equine infectious anemia virus
FeLV	feline leukemia virus
HERV	human endogeneous retrovirus
HFV	human foamy (spumaretro) virus
HIV	human immunodeficiency virus
HTLV	human T-cell leukemia virus
IAP	intracisternal A-particles
IN	integration protein
IPTG	isopropyl thiogalactopyranoside
LAV	Lymphadenopathy associated virus
MA	matrix protein
MES	morpholinoethane sulfonic acid
MMTV	mouse mammary tumor virus
MNM	1-deoxymannojirimycin
MoMLV	Moloney murine leukemia virus
MPMV	Mason-Pfizer monkey virus
M-RadLV	murine radiation leukemia virus
MSV	murine sarcoma virus
MuIAP	murine intracisternal A-particles
MuLV	(Moloney) murine leukemia virus
NC	nucleocapsid protein
ORF	open/overlapping reading frame
PAGE	polyacrylamide gel electrophoresis
PCMB	4-chloromercuribenzoic acid
PCMS	p-chloromercuriphenylsulfonic acid
PIPES	1,4-piperazinediethanesulfonic acid
PMSF	phenylmethylsulfonyl fluoride
PR	protease
PVDF	polyvinylidene difluoride
RNase H	ribonuclease H
RP-HPLC	reverse-phase high pressure liquid chromatography
RT	reverse transcriptase
SDS	sodium dodecyl sulfate

SIV	simian immunodeficiency virus
SRV	simian retrovirus
STI	soybean trypsin inhibitor
STLV	simian T-cell leukemia virus
SU	surface protein
TLC	thin layer chromatography
TM	transmembrane protein
VLV	visna lentivirus

SUBJECT INDEX

Bei Fragen zur Produktsicherheit wenden Sie sich bitte an:
If you have any questions regarding product safety,
please contact:

Walter de Gruyter GmbH
Genthiner Straße 13
10785 Berlin
productsafety@degruyterbrill.com